ICT and Primary Science

ICT is often used as a quick way of recording data in the primary science classroom, but when ICT becomes an integral part of the primary science, both subjects become more exciting, meaningful and relevant.

Throughout this book, the authors emphasise that primary science is at its best as a practical, hands-on experience for children. When ICT is used in an integral way, it can enable practical work to be done at a more sophisticated level, helping children to make sense of their findings. The book includes several case studies from primary classrooms, and each chapter includes practical suggestions for teachers.

Topics covered include:

- Databases
- Spreadsheets
- Data logging
- Control technology
- ICT, drama and science
- School visits
- Planning for ICT and science
- Choosing and using software

ICT and Primary Science is an accessible and jargon-free resource for teachers and student teachers of primary science.

John Williams has recently retired from full-time teaching at Anglia Polytechnic University. He remains involved with primary education, lecturing, supervising student teachers in schools and writing. **Nick Easingwood** is Senior Lecturer for ICT in Education and acts as the ICT Co-ordinator for the School of Education at Anglia Polytechnic University.

ICT and Primary Science

John Williams and
Nick Easingwood

Routledge
Taylor & Francis Group
LONDON AND NEW YORK

First published 2003 by RoutledgeFalmer

This edition published 2013 by Routledge
2 Park Square, Milton Park, Abingdon, Oxfordshire OX14 4RN
711 Third Avenue, New York, NY 10017

First issued in hardback 2017

Routledge is an imprint of the Taylor & Francis Group, an informa business

Typeset in Melior by
HWA Text and Data Management, Tunbridge Wells

British Library Cataloguing in Publication Data
A catalogue record for this book is available from the British Library

Library of Congress Cataloging in Publication Data
A catalog record for this book has been requested

ISBN 13: 978-1-138-16638-7 (hbk)
ISBN 13: 978-0-415-26954-4 (pbk)

Contents

	List of illustrations	vi
	About the authors	viii
	Foreword	ix
	Acknowledgements	xi
	Introduction	1
Chapter 1	Planning, organising and managing ICT for the teaching of science	11
Chapter 2	Selecting software for science	31
Chapter 3	Databases	49
Chapter 4	Spreadsheets	71
Chapter 5	Data logging	81
Chapter 6	Control technology	95
Chapter 7	Museums, zoos, gardens, art galleries and ICT	119
Chapter 8	Science, ICT and drama	135
Chapter 9	Using ICT to present the findings of scientific investigations	143
	Index	159

List of illustrations

1.1	Example of a lesson plan detailing the woodlice experiment.	13
1.2	*Number Box* spreadsheet that records the migration of woodlice from light to dark conditions every minute.	15
1.3	Line graph to illustrate the migration of woodlice from light to dark conditions.	16
1.4	Column chart.	16
1.5	Bar chart.	17
2.1	*Number Box*: Yellow level.	41
2.2	*Number Box*: Red level.	41
2.3	*2Simple Video Toolbox: 2Count.*	42
3.1	Scattergraph produced by *Information Workshop*.	52
3.2	Step 1.	57
3.3	Step 2:	57
3.4	Step 3.	58
3.5	Step 4.	58
3.6	Step 5.	59
3.7	Step 6.	59
3.8	Step 7.	60
3.9	Card used by the children to record their data.	62
3.10	Table produced in *Information Workshop* containing all of the information entered by the children.	63
3.11	Table showing animals sorted by their habitat.	64
3.12	Graph illustrating the distribution of animals in the survey around the world.	65
3.13	3-D pie chart illustrating the same information as Figure 3.12.	65
3.14	Information sorted and displayed by diet.	66
3.15	Pictures created using the 'Faces' screen of *My World.*	67
3.16	Record card completed with a pencil.	68
3.17	Record card produced using *Information Workshop*.	68
4.1	*Number Box*: 'Temperature' quick sheets.	74
4.2	'Pulse Rates' quick sheets.	74
4.3	'Growing a Plant' quick sheet.	75
4.4	Spreadsheet used to record the number of ruler drops.	77
4.5	Graph showing each of the children's individual reactions.	78
4.6	Graph showing an average reading for each of the children's reactions.	78
5.1	EcoLog from Data Harvest.	86
5.2	EcoLog with temperature and light sensors.	87
5.3	EcoLog: Two temperature sensors logging warm and cold water.	88
5.4	An impression of what the graph of the temperatures and sound levels from inside one of the trees might look like.	89
6.1	Control box and model.	98
6.2	Side view (not to scale) of a gear train frame.	101

6.3 Plan view of a gear train (not to scale). 102
6.4 A gear train. 103
6.5 A gear train *in situ*. 103
6.6 Close up of a gear train *in situ*. 104
6.7 The gear train is connected to a set of cams which are used for alternating the movement of the sculptures fixed to the top of the pistons. 104
6.8 The gear train fixed to the cams allows for the cat's mouth to open and close. 105
6.9 Close up of cams attached to a gear train. 106
6.10 Two children using the Ladybird. 110
6.11 Drawing of a Ladybird produced by a member of Willow Class. 111
6.12 Another drawing of a Ladybird produced by a member of Willow Class. 111
6.13 Cover designed by Amy to illustrate her robotics work. 113
6.14 Writing about the robot. 113
6.15 Writing about programming the robot. 114
6.16 Describing how the robot was programmed. 114
6.17 Describing the Ghost Train. 115
6.18 Two children's descriptions of programming the Ghost Train. 115
7.1 Example of a worksheet for a gallery in a museum on the theme of 'Flight'. 122
7.2 Diagram to show the balance of forces that keep an aeroplane in flight. 124
7.3 Diagram to show how the ailerons could be positioned. 124
7.4 Type 1 lever, such as is found on a see-saw, or when lifting a heavy object. 128
7.5 Type 2 lever, such as that found on a wheelbarrow. 128
7.6 Type 3 lever, such as that found on the forearm. 128
7.7 Archimedes' screw. 129
7.8 Try different-sized cylinders with differently spaced tubes. 130
9.1 Printed text at ×10 magnification. 147
9.2 Printed text at ×60 magnification. 147
9.3 Printed text at ×200 magnification. 147
9.4 Instructions describing how to make a lava lamp. 153
9.5 An acrostic poem about fire. 153

About the authors

John Williams has worked in Primary Education for over thirty years as Class Teacher, Head Teacher, Advisor for Science, Governor and as a Senior Lecturer in Higher Education. It was during his time as a Head Teacher in Kent that he realised the potential of ICT in education, and his school was one of the first in the country to use computers for teaching primary age children. He has written many books on primary science, and Design and Technology. He has recently retired from full-time teaching at Anglia Polytechnic University, where he was the Admissions Tutor and a lecturer responsible for the teaching of Design and Technology to student teachers. By dividing his time between Italy and the UK, he remains involved with primary education, lecturing, supervising students in schools, and of course more writing.

Nick Easingwood is Senior Lecturer for ICT in Education and acts as the ICT Co-ordinator for the School of Education of Anglia Polytechnic University. Apart from leading a Post Graduate Certificate of Education (PGCE) Secondary ICT course, he also contributes to Primary and other Secondary Initial Teacher Education courses, as well as in-service Bachelors and Masters degree courses. He maintains regular contact with schools through visiting students on their school experiences, having himself spent 11 years as an Essex primary school teacher. His research interests include the development of online resources for student teachers and their capabilities in ICT on entering Higher Education. He has a growing publications record, including *ICT and Literacy* (2001) which he jointly edited, as well as a contribution to the second edition of *Beginning Teaching, Beginning Learning*, edited by Janet Moyles and Gill Robinson, which was published in April 2002.

Foreword

Since my own Primary School youth nearly 70 years ago, or even that of my sons, there has been a revolution. As a physicist, I could describe it as a massive increase in the spectral width of the material, which is presented to the children. I have deliberately avoided the use of the word 'taught', because this is not quite how it is done, it much more involves a variety of measures to engage the interest and imagination and to make the experience more relevant to the outside world. This book deals specifically with the application of ICT to Primary Science, though as the authors admit, it does of course spill over into other branches of the curriculum.

When I first considered writing this Foreword, the thought occurred to me that in the world of ICT it was possible for even quite young children to achieve a greater degree of competence than their parents. Was that a novel situation? It probably is not, because there must have been similar transitions when children who could read had parents who could not, or later when Science was taught, there were children who could demystify everyday events using scientific principles. I remember discussing with John Williams how one could explain the phenomenon of 'floating', when clearly the materials from which the boat was made did not float, a phenomenon, which very likely is still opaque to many adults.

A possibly unintentional side-effect of the all-pervasive use of ICT in Primary Science teaching, is the 'trickle up' from the children to their parents.

What this book advocates is that the use of ICT in the assimilation and investigation of Science should become as natural and matter of course as the use of pencil and paper and a stop watch, for a previous generation. The fact that the 'piece of paper' can do sums and plot graphs, and store data in a systematic form, just to mention some of the facilities, enormously expands the capabilities of what a child could hope to achieve in a given time-frame.

The advocacy takes the form of teaching the teacher, by explaining the function of different software programs that are available. Fortunately, these are now much more like those used in the real world than was formally the case and make the transition to adult practices, or even what Mum and Dad are doing, that much easier.

The approach is hands-on, specifying what is needed and suggesting experiments to perform.

There is also a chapter on Control Technology, representing a very important branch of ICT, which is not often taught. The authors go to considerable trouble to guide the reader into ways of presenting the problem of automatic control to the children, by getting them to invent the algorithm and then implementing the overall system with hardware which is easy to obtain.

The merit of the book is that it is thorough and illuminates the subject from many different directions. It also functions as a complete 'how to' guide, without ever making demands on resources, which are unlikely to be relatively readily available. The fact that the authors also reiterate the cross-curricular nature of the application of ICT in a Primary School setting, emphasises correctly that ICT is now firmly established in our culture, just like reading and writing!

This book makes an important contribution not only to Science teaching but also to the cultural assimilation of ICT.

Prof. Heinz Wolff
Brunel University

Acknowledgements

The authors would like to thank the following companies, schools, organisations and individuals for their support and help in the writing of this book.

Suppliers

2Simple Software, 3-4 Sentinel Square, Brent Street, Hendon, London NW4 2EL
Commotion Group, Unit 11. Tannery Road, Tonbridge, Kent TN9 1RF
Data Harvest Group Limited, Woburn Lodge, Waterloo Road, Linslade, Leighton Buzzard, Bedfordshire LU7 7NR
Granada Learning Ltd (Black Cat), Granada Television, Quay Street, Manchester M60 9EA
Lego Robolab, Unit 11. Tannery Road, Tonbridge, Kent TN9 1RF
Microsoft UK, Microsoft Campus, Thames Valley Park, Reading RG6 1WG
TTS Limited, Monk Road, Alfreton, Derbyshire DE55 7RL

Schools

The authors would particularly like to thank the staff and children of the following schools for the help, co-operation and hospitality shown to them during the writing of this book.

Elaine Primary School, Elaine Avenue, Strood, Rochester, Kent ME2 2YN
Bluehouse Community Junior School, Leinster Road, Laindon, Basildon, Essex
St. Christopher School, Letchworth, Hertfordshire SG6 3JZ
Tewin Cowper Endowed Primary School, Cannons Meadow, Tewinl, Welwyn, Hertfordshire AL6 0JU
Westbury Primary and Nursery School, West View, Letchworth, Herts SG6 3QN

Other addresses

Association for Science Education, College Lane, Hatfield, Hertfordshire AL10 9AA
British Society for the History of Science, 31 High Street, Stanford in the Vale, Faringdon, Oxfordshire SN7 8LH
Primary Schools Development Executive, ECB Education Department, Lancashire C.C.C. Indoor Cricket Centre, Old Trafford, Manchester M16 0PX.

Our thanks are also extended to Initial Teacher Training students of the School of Education of Anglia Polytechnic University, Chelmsford, Essex, especially the following students who kindly allowed the authors to use images of their models in the Control Technology chapter and their PowerPoint presentations on the accompanying website: Hayley Blyth, Claire Chick, Julie Fawthrop, Clare Frostick, Victoria Hawkins, Clare Johnson, Jodie Lawrence, Jodie Lock, Katy McKee, Toni Paul, Lea Tyler and also to staff colleagues Mark Miller, Alan Myers and Grace Woodford.

Screenshots and Clip Art reprinted by permission from Microsoft Corporation. Microsoft is a trademark of the Microsoft Corporation.

Line drawings 5.4, 6.2, 6.3, 7.4–7.8 by John Williams.
Illustration 7.1 by Nick Easingwood and John Williams.
Illustrations 7.2 and 7.3 by John Williams and Mark Miller.
All photographs by Nick Easingwood except for 6.10, which is courtesy of Westbury School, Letchworth.
All other illustrations by members of the schools detailed above.

Figures from suppliers

2Simple (2.3)
Granada – Black Cat (1.2–1.5, 2.1, 2.2, 3.1–3.14, 3.17, 4.1–4.6)
Granada – My World (3.15)
Commotion Group (6.1)
Data Harvest (5.1–5.3)
Intel Play and Mattel (9.1–9.3)

Introduction

> We can learn a lot from children, and especially from watching children think. Children can be brilliant thinkers.
>
> Edward de Bono, *Children Solve Problems* (1972)

It was in 1989 when the first National Curriculum documents were published. For science, the document was very comprehensive, containing seventeen 'Attainment Targets', each representing different areas of the subject, including its history. The fact that it was one of the first subject documents to be produced, to some extent indicated the importance given to science within the curriculum at that time. However, little if anything was said about the use of ICT.

Things are very different now. The 1989 document was undoubtedly just too detailed, but in its several revisions, it can be argued that much of what it gave to science teaching: imagination, breadth, historical background, has been lost. Nevertheless the present document still allows schools, and the individual teacher, plenty of scope for science investigations to be carried out in some detail. The handbook alone gives details of the links that exist with other subjects, and even at Key Stage 1, refers to complete investigations. Moreover, although the overall content has been reduced, it now includes many references to the use of ICT, as well as this subject now having its own separate document.

However, perhaps because of 'League Tables' or because of the introduction of SATS, there is a danger that primary science could become a watered-down version of the old lesson-based subject, which had little continuity, and even less relevance for the pupils, or to the world in which they live. Nevertheless, science remains a core subject, and therefore retains its place in the increasingly crowded primary school curriculum.

In this book, the authors have considered various aspects of science teaching, and how they can be enhanced by the appropriate use of relevant ICT. One of the beneficial changes that has taken place since the science national curriculum was first published, is the large-scale introduction of ICT, not just in science, but into all the other subjects as well. The present science curriculum document not only gives references to links across the curriculum, but also specifically requires teachers to use ICT in the collection and collating of data. Moreover, it also requires the teaching of such techniques as data logging, and the use of CD-ROMs and the Internet.

ICT has its own unique place both within the general school curriculum, as well as specifically within the National Curriculum itself. However, there is now a move towards emphasising the use of ICT as an integral part of each and every subject. It is within science that arguably one of the closest links exists. In their recent progress report, OFSTED highlighted the fact that many teachers remain in doubt as to when and when not to use ICT. This book will not only help teachers to understand when to use ICT, but also what kind of ICT is most appropriate, and at what stage. At the same time (the authors make no apologies for repeating this message later), ICT needs to be used to enhance what must in the first place be good science. To make the best use of ICT it will be necessary to carry out detailed scientific studies, whether they include ICT from the start, such as when using data-logging equipment, or later when collating the results gained during the investigation with the help of a spreadsheet or database.

Whilst we have included two chapters on planning and the selection of software, most chapters concern the specific areas of ICT with which most teachers are familiar, such as databases, spreadsheets or control technology. We have purposely included, usually at the start of the chapter, some background information, so as to put into context the general educational value of the particular aspect of ICT being discussed. This may involve some historical detail to show how ICT has been able to enhance an evolving science curriculum. In some chapters, such as the one on databases, it has been necessary to give quite detailed definitions. Databases and similar applications can be very complex, and there is much discussion about how many variations there are, and just what constitutes a database anyway. We trust that we have been able to simplify this to a great extent. It is our intention that this book should be more than a handbook for teachers that just includes a check-list of ideas or a series of simple 'instructions for the classroom'. We hope it proves to be a little more thought provoking than that! Nevertheless we have included many detailed practical suggestions, both for the teaching of science and the application of ICT. Where possible, we have included one or more case studies within the chapter. These show how teachers in a wide range of schools are able to teach these projects in situations that most of the readers of this book will easily recognise, especially in the light of the completion of the ICT training funded by the New Opportunities Fund (NOF). The final three chapters cover wider aspects of the use of ICT in primary science, which despite the present somewhat restricted curriculum, are still important areas of study for many primary teachers and schools.

It is for this reason that this book is also more than a manual on how to 'use' ICT. It is hoped that it will give as much help to teachers of science as well as those of ICT. All the science described can be found within Key Stages 1 and 2 of the National Curriculum. Where necessary it is specifically labelled as such, but other examples are nevertheless within the statutory requirements, and can be placed there with a little careful reference to the National Curriculum. Whilst we have of course made reference to other sources and texts, all of which are acknowledged either specifically or in the bibliography at the end of each chapter, most of the content has been drawn from our own experiences. The authors have both spent a considerable part of their professional lives teaching primary school children of all ages, and it is hoped that we can be forgiven for allowing this experience to be reflected in the text!

Although science remains a core subject, due to the many pressures on the timetable, schools have found it difficult to allot it the amount of time that it once enjoyed. Consequently, despite the requirements of Sc1, some of the basic tenents of the subject can be lost. It needs to be remembered that the basic skills of science still need to be included in any useful lesson. Most, if not all science should always start with the process of observation. This will pose questions such as: What is there? Why is this there? What is happening? The answers to these questions can then be analysed, and hypotheses postulated. These hypotheses can then be tested by experimentation, and decisions made depending upon the results. These in turn need to be recorded, with probable further analysis. All these skills and processes fall into the Knowledge, Skills, and Understanding and Scientific Enquiry areas of the National Curriculum. If there is a technological element in the science, which in the primary school could for example be the making of musical instruments (sound or materials), or making an electrically driven model or a reflective toy, then the communication of the experimental results can lead to really useful and relevant Design and Technology.

Naturally there are occasions when this step-by-step approach to scientific enquiry can be too restricting. At Key Stage 1, just simply watching ice melt, and asking for ideas from the children about this observable change of state, is a valuable exercise in itself. It should not at this stage require much, if any detailed recording. Later, at Key Stage 2, if this same exercise could be carried out in the way that Faraday did it, then it would be very valuable indeed, and would fit within the Materials and Their Properties part of the National Curriculum. Too slavish attention to the 'scientific process' can also delay the final discovery. Although it is fairly argued that Rosalind Franklin should have shared the Nobel Prize for the discovery of the structure of DNA, Watson and Crick, by not following the more structured experimental approach favoured by Franklin, won the prize. They made a conceptual leap, in part based on some of Franklin's own work, and so were able to construct the first model of a DNA molecule. (Incidentally, using simple constructional techniques recognisable to most primary school teachers.)

There are times, therefore, when an inspirational idea can allow the scientist to leapfrog several of these experimental stages. The history of science is littered with such examples, although they were usually based on a sound knowledge of the subject as it was understood at the time. If this should occur during a scientific enquiry in the primary school, then an imaginative and knowledgeable teacher would make the necessary allowances. Teachers, however, also need to be aware that the results of the experiments may simply be wrong, or at least doubtful, and will need to be questioned. There is also the ever-present temptation to make the results fit the hypotheses, despite evidence to the contrary. ICT can help with these problems, both as a source of information, and as a means of communication.

Since the election of the first Labour Government in 1997, much has been made of the place and purpose of ICT in British schools. What is all the fuss about? Why have several hundred million pounds been spent on providing schools with computers, connecting them to the Internet and training the teachers how to use them? Why can the use of ICT enhance the teaching and learning, and in particular, science in the primary school? How can the use of ICT enable access to new levels of learning and develop and extend teaching? The main purpose of this book will be to answer these key questions. Modern hardware and software brings new resources and advantages that were unavailable to teachers even three years ago. The advent of the wide-scale adoption of ICT suites in primary schools, their connection to the Internet and Email via the National Grid for Learning and the availability of relatively inexpensive ICT peripherals such as digital cameras, scanners and electronic microscopes, provide teachers and learners alike with a new armoury of tools that are powerful yet very easy to use. For what is captured can be instantly and easily used in a wide range of applications and contexts, all of which can enhance the teaching and learning of primary science.

When analysing the contribution that ICT can make to primary science, or indeed any subject that is part of the curriculum, it is very important to think of ICT in an integrated, holistic way. ICT is not just about the computer. It is also about the ICT suites, the software, the hardware, the networks, the peripherals as well as other resources such as television, video and tape recorders. Above all else, it is about the subject and the user. If ICT is to be used successfully to support science then it is important for the teacher and ultimately the pupil to appreciate that ICT is not about teaching and learning computer skills, or even about 'being good at computers'; it is about employing ICT to support the development of subject-specific skills. This may, in some cases require a fundamental philosophical shift on the part of the teacher. ICT can be used to support all of the basic scientific skills, as well as the more modern aspects of research and communication (which was why the 'C' of ICT was added in the first place) through the use of CD-ROMs, the Internet and email. However, in any investigation, the science should always come first, and the ICT second. Obviously there will need to be some ICT skills development in order for the user to use the equipment effectively, for we want the keyboard to act as a gateway to what is contained beyond, not as a barrier. This book will seek to illustrate how this access can be easily attained, quickly and with the minimum of ICT skills development. It is intended that the necessary ICT skills will be developed simultaneously with the science skills.

So what can ICT bring to primary science? A modern computer system can handle huge amounts of different types of data and information quickly, automatically and in an integrated and provisional way. It can run a word processor, presentation software, a desktop publishing package, a spreadsheet, a database, and it can use a graphics package. Through the use of a scanner, digital camera and appropriate software, the computer can capture and alter images that in turn can be cut or copied into a wide range of other applications. The computer can also be used as a means of accessing the content of the Internet, where pages can be viewed, copied or saved for later use in a variety of different ways. It also provides the means for the sending and receiving of email. Some of the files and documents produced in the applications listed above can in turn be attached and sent to distant parts of the country or the world. It can log environmental data and run CD-ROM programs that contain factual information, games or simulations. It can be used for video conferencing. It can run a web cam so that users can communicate and see each other over the

Internet. Any information that is created and saved can be printed or saved to a CD-ROM, a floppy disk or a high capacity zip disk. It can be used for the design and posting of web pages onto the Internet. It is able to program and to control models, a robot, or a floor turtle. Any one of these uses can be integrated with any other through the use of copying and pasting. A document that has been produced in a word processing package can be pasted into a web page, a newsletter or sent as an email attachment. The findings of an investigation can be displayed as part of a wider presentation. All of this can be done quickly, easily and automatically. This provisionality means that not only can work be reformatted for use in different applications or ways, it can also be easily altered. Any work that is produced is never completely finished. A piece of work completed with traditional pencil and paper, eventually reaches a conclusion. Indeed, to do any more to it can often spoil or even ruin it. However, when ICT is used, what previously might have been the end of the design and creation process can now be the beginning. Above all, it allows for genuine pupil interaction, whereby the user is not a passive recipient of the material on the screen. The use of ICT allows the user to create, refine, modify, alter and edit work in a wide number of ways. This is a powerful aspect of ICT that should not be overlooked.

The National Curriculum for England (DfES, 1999) has many references to ICT contained within it, and the science document details specific ICT opportunities on most pages. The programme of study for Science at Key Stage 1 emphasises the fact that under investigative skills, ICT can be used for 'obtaining and presenting evidence, specifically placing it in the context of communicating through the use of speech and writing, drawings, tables and block graphs' (Sc1: 2g). Under 'Breadth of Study' (1c) there is a requirement to 'use a range of sources of information and data'. In 'Scientific Enquiry' at Key Stage 2 there is specific mention of data logging – the only direct reference to a specific application (2f).

The ICT programmes of study are geared very much towards the idea of ICT as a tool for learning rather than the need to teach specific ICT as such. For example, to look at just one of the many requirements within these programmes of study, the 'Knowledge, skills and understanding' section of Key Stage 1, which describes the need to 'gather information from a variety of sources' (1a), 'enter and store information in a variety of forms' (1b), 'use text, tables, images and sound to develop their ideas' (2a), to 'share their ideas by presenting information in a variety of forms' (3a), and to 'review what they have done to help them develop their ideas' (4a), shows us that the ICT should be an integral part of other subjects. All aspects of the programmes of study for ICT can be easily covered whilst teaching primary science.

However, although a legally binding document in England, the National Curriculum is a minimum requirement. The British Educational Communications and Technology Agency (BECTa) describe other, more specific reasons for using ICT in the teaching of science. According to their excellent web site:

> ICT can enhance teaching and learning in science. Science also provides a context in which pupils can develop ICT capability and a broader 'technological literacy'. ICT can:
>
> - facilitate links with other places, subjects and other people
> - facilitate the asking of questions and forming or modification of opinions
> - provide access to secondary sources of information with more breadth and depth
> - support communication, thereby raising issues of audience and viewpoint
> - enable the gathering, storage and manipulation of data and other information
> - enable more effective analysis of data and information
> - enable the simplification, simulation and modelling of scientific ideas
> - enable more effective communication of understanding or experimental results
> - support the asking of 'What if...?' questions through experimentation and testing
> - support teachers' professional development alongside pupils' learning.

It goes on to say:

> The use of ICT in school science falls into two broad areas: communicating and data handling. These can be broken down into more recognisable classroom activities:
>
> - Researching – real-time experimenting with, for example, data loggers
> - Searching for information – from the Internet, e-mail and CD-ROMs
> - Analysing data
> - Simulations/demonstrations
> - Modelling
> - Drawing diagrams, writing up and presenting findings.

http://www.becta.org.uk/technology/software/curriculum/reports/science.html (accessed June 2002)

Indeed, the careful and selective use of ICT can help with all the processes of science. Moreover, it can shorten the time it takes to record and produce data. If whole mornings are to continue to be taken up with the literacy and numeracy element of the timetable, then the more we use ICT as a properly integrated part of the science lesson the better. If the wonder and excitement of science is not to be lost, then it is imperative that ICT should be included, not just to streamline the process, but to motivate the children and enhance the quality of their learning. The processes of science may change or become modified with time, but the basic skills remain, although alongside them there now has to be added a new set associated with the use of ICT.

Unlike the processes of scientific discovery, one of the peculiarities of science education is that the science component often lags behind the educational processes. It is not simply that when the science reaches the school it is out of date, but that the progress of scientific discovery has become so rapid. This has probably always been the case. Fifty years ago, when browsing through an atlas, pupils would have likely as not been told that although it seemed that the land masses could be made to fit together like the pieces in a jigsaw puzzle, this was purely coincidental. Nevertheless, at about this time scientists were discovering the phenomena of plate tectonics and continental drift.

In this book we have not attempted to measure the rate of scientific progress. Nevertheless, we would not be surprised to find that today this progress increases exponentially. In other words, scientific discovery feeds upon itself, so that any single advance adds to the increasing pace of scientific development. Consider the history of flight. It was only at the turn of the last century that powered flight was possible. Since then, in the relatively short time-span of one hundred years, we have invented the jet engine, flown supersonically, reached the moon and have developed aircraft that can carry over 500 passengers to any part of the world. In an even shorter time-span, because of the technological advances in astronomy, we now recognise that there are billions of galaxies, where previously it was thought that there was only one.

Although the development of ICT may today be even more rapid than science, the two are now inexorably linked. They are mutually dependent in that ICT assists scientific development and vice versa. Indeed, if a similar situation arose to the plate tectonics example above, we would expect that through the help of ICT the teacher and pupil would reach the same conclusion. We hope that in this book we have shown that even with the youngest children there should be an inseparable connection between science and ICT. The case studies give examples of how programs, which are easy to use, enhance both the science and the educational experiences of the children. We have also suggested many other uses of ICT and science that rely on a range of software that is readily available in schools or on the World Wide Web. All that is needed is for the available software to be used to its fullest potential. We have given no examples of science or ICT that are beyond the scope of all primary schools. Although there needs to be a willingness to explore the software's full capabilities, and to assess their value to science education, what is required for their effective use is a parallel appreciation of a true understanding of primary science, together with a knowledge of the place and purpose that ICT can play in its teaching.

We are very aware that hard-pressed and over-worked teachers may find it difficult to devote the time to incorporate these ideas into their teaching. However, we believe that this is so important that it is inevitable that they will be allowed this time and this has to be reflected by a corresponding decrease in their bureaucratic workload.

Bibliography

De Bono, E., *Children Solve Problems*, Allen Lane, Penguin Education, 1972.
Madox, B., *Rosalind Franklin, a Biography*, HarperCollins, 2002.
DfES, *The National Curriculum, Handbook for Primary Teachers, Key Stages 1 and 2*, DFEE, 1999.
OFSTED, *ICT in Schools, A Progress Report*, (HMI 4, 23), OFSTED, 2002.
Wenham, M., *Understanding Primary Science: Ideas, Concepts, and Explanations*, Paul Chapman Publishing, 1995.

http://www.becta.org.uk/technology/software/curriculum/reports/science.html (accessed June 2002).

Chapter 1

Planning, organising and managing ICT for the teaching of science

The first and most crucial point to make about the use of ICT in the teaching of science is that it cannot and will not replace the teacher. Computers have many assets, all of which can be used to assist and enhance the teaching of science. However, computers are only boxes of electronics – they cannot, as yet, think for themselves, get to know each one of their pupils on a personal level, engage in conversation about their interests, plan, prepare and assess their work on an individual basis or interact with their parents at open evenings. The important fact to remember is that good teaching and learning can only occur when a good teacher is present – and that a good teacher is not a computer. The key to successful teaching and learning with ICT lies in how the technology is used and employed, not in the teaching of the technology itself. It is this that makes the teacher's role so crucial. In the introduction we detailed *why* ICT should be used in the teaching of primary science. This chapter will detail *how* in terms of the practicalities that are involved in planning, assessing, organising and managing teaching and learning where ICT is involved.

In order to illustrate this clearly to the reader, we begin this chapter with a case study that details how ICT is used to support a practical science lesson. Following an explanation of the nature and purpose of the investigation, a sample lesson plan is provided to show the reader how such a lesson might be planned. A series of screenshots illustrate the kind of findings that might be expected from this investigation where ICT has been used. These are discussed with the 'whole lesson' being broken down into the constituent parts and considered in turn. This will provide a clear explanation of what a lesson where good ICT is used might look like.

Case study – woodlice

The woodlouse has always been an interesting animal to study. It is relatively large, easy to find and has an interesting biological history. Some years ago, many Primary Schools were using a project box, which included a range of materials from large-scale card models of the animal to detailed drawings and descriptions of its anatomy and physiology. Perhaps it is time for another look.

Wherever possible, animals should be studied in their natural habitat. Obviously, this is not possible in most cases, and we are certainly not suggesting that because children cannot be taken to India they should not learn about the tiger, any more than they should not learn about tropical fruits because they do not grow naturally in this country. It has always been one of the strengths of primary teachers that they were able to devise natural and relevant situations to bring to children things that would normally be outside their experience. Visits to zoos, gardens, museums, and now various uses of ICT are obvious ways that can help children make sense of the wider world. Even sport has contributed to a wider learning. Football teams and geography combine very well, and the recent project pack for Key Stage 2, *Howzat!* produced by the ECB and Channel Four does much the same thing. This pack, although designed to encourage interest in cricket, takes a very cross-curricular approach, and includes worksheets for both science and ICT.

One great advantage with the woodlouse is that whilst it can be studied in its own habitat, it also makes a very good classroom animal. Woodlice can be kept in glass tanks, and all they need is a little damp soil, leaf mould, some rotting wood, and a stone or two. However, because they are so abundant, woodlice can easily be studied in their natural habitat. Indeed they can often be found in the most urban of habitats. They need only be brought into the classroom for the duration of the experiment, and after it is complete they can be returned to exactly where they were discovered. The experiment detailed below takes the form of a simple choice chamber, the animals having to 'choose' between two sets of conditions. Obviously, some of these conditions may already be apparent, particularly to the children. If all the woodlice were found under stones, then it may seem obvious that they prefer dark, damp conditions. This does not matter, not only because some children might suggest that the woodlice were sheltering from the approaching feet – a good point to make – but also because the main aim of the lesson is to teach experimental techniques. Learning how to plan and carrying out a fair test, and collect and collate the data, and its subsequent interpretation, are all essential in themselves. The fact that at the same time, this will also teach the children more about the living world, makes it all the more realistic and valuable.

Figure 1.1 shows a lesson plan that could be used to teach a lesson involving woodlice and their habitat. The numbers in parentheses act as part of a sub-heading for the discussion that follows.

(1) Curriculum Area: Science

Focus: To study behaviour of woodlice in their habitat and in classroom conditions.

(2) Objectives of Lesson – NC Refs. Sc1/2, Sc2 4a, b, c; 5a, b, c.

- To understand what kind of animal it is.
- To understand something about its natural environment.
- To learn how to undertake a 'fair test' in the classroom in order to discover something of it's preferred habitat.
- To learn to take care of, and have consideration for the well being of all types of animal.

(3) Prior Learning – General study of animal types, with particular focus on 'minibeasts'.

(4) Development of Lesson

Children will be reminded about the woodlice from their previous work with 'minibeasts'.
The class will then be taken into the school ground (garden or wild area), to look for woodlice.
They will take great care when collecting them not to harm them in any way.
When the woodlice are brought back into the classroom, they will be distributed equally within the groups. Each group will carry out a set of experiments-
Woodlice placed in a plastic tray. The tray will be two parts, each part being a different and opposite habitat – i.e. light and dark, warm and cool, dry and damp. Each tray will have ten woodlice, and the children will record the position of the woodlice, over a period of 15 minutes.

On return to the classroom they will enter their findings into a spreadsheet using *Number Box*, part of the *Black Cat Toolbox,* and produce a range of graphs to illustrate the movement from light to dark.

(5) Teaching Points

They will use books, pictures, and a CD-ROM.

The children will be encouraged to take note of where these animals can easily be found i.e. under stones, in rotting wood.
The children will be encouraged to treat these animals with the same care, as they would give to their pets at home. They are all living creatures.
There will be six mixed ability groups.
One person per group will handle the woodlice. Others will have the tasks of counting, timing, and recording. All will be involved in setting up the experiment, and changing and designing the range of habitats. They will be encouraged to think of different kinds. After completion, the woodlice will be returned to where they were found. All children will wash their hands and clean equipment.
Children to look for patterns –

- What seems to have happened?
- Why do they think that this has happened?

(6) Differentiation

During the collection of the woodlice, and the setting up of the experiments there should be little need for differentiation. However, for the reinforcement period, and the subsequent recording, special designated worksheets, and a CD-ROM will be made available.

continued overleaf

(7) Assessment
Can the children appreciate the concept of a fair test?
Why is a fair test necessary in this instance?
Why have the woodlice migrated?
What is the graph showing?

(8) Cross-curricular Links

Introduction/reinforcement of negative numbers (Mathematics)
Introduction/reinforcement of line graphs
Selecting appropriate graphical representation

(9) Resources

Identification books, CD-ROM, Pencils, ClipBoards, Collecting Jars, Grey Plastic Trays, Black Paper, Light (torches?) for warmth experiment, Sand, Clean Garden Soil, Fine Gravel. (A variety of similar materials should be available, as the children will be asked for their ideas about differing habitats). Timers or Stopwatches, Graph or Squared Paper. Water should also be available. Access to appropriate web sites.

(10) Follow-up

Use of ICT:

- Spreadsheet package (*Number Box* from *Black Cat*) to illustrate migration of woodlice from light to dark
- Branching Database to classify the woodlouse within the animal kingdom (Note: the woodlouse is our only terrestrial crustacean hence its requirement for a damp environment. It has evolved from an aquatic environment and still needs some remnant of this habitat).
- CD-ROMs and web sites to use as a source of research about minibeasts.

Figure 1.1 Example of a lesson plan detailing the woodlice experiment.

Combining the science with the ICT

The lesson plan in Figure 1.1 describes the investigative phase of the science activity. Although some ICT is considered to assist with researching woodlice, the focus here is very much upon the practical aspects of 'hands-on' traditional, primary school science. In the last section, detailing the follow-up work, the ICT has greater focus, as this is where the raw data that has been gathered by the class is then processed.

The teacher can decide to show the results in various ways, graphs, spreadsheets, written tables, even pictures may all be appropriate, and can depend on the abilities of the children. The examples shown here are taken from work carried out by children in a Year 6 class some years ago. One of the authors was the class-teacher at the time. The school, Elaine Junior School, Strood, in

Kent, (as it was then) had no computers, for this work was undertaken before they became available. All the work was recorded on paper. However, it is used here to show how work of this kind fits well with both the present National Curriculum, and with the requirements of ICT. The graphs used had the *x*-axis (time) in the 'middle' of the *y*-axis (number of woodlice). The subsequent graphs should of course be mirror images of each other, and the numbers below the *x*-axis are in effect minus numbers. They reflect what is not present above the *x*-axis. As has been previously explained, the fact that the children might have anticipated the result does not matter. They could always carry out the experiment again, this time putting the woodlice in the opposite habitat. The subsequent results will at best confirm the experiment, or at the worst provide a lot of discussion! Some possible outcomes are illustrated in Figures 1.2 to 1.5.

Figure 1.2 shows a Number Box spreadsheet that records the migration of woodlice from light to dark conditions every minute. The minus value on the number is necessary to display this graphically, as illustrated by the line graph in Figure 1.3.

Alternatively, the information can be displayed as a column chart, as shown in Figure 1.4 or as a bar chart, as in Figure 1.5.

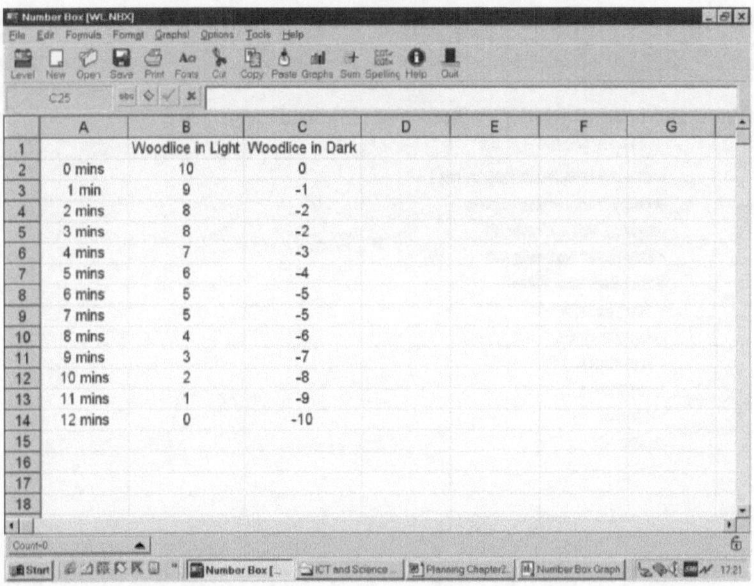

	A	B	C
1		Woodlice in Light	Woodlice in Dark
2	0 mins	10	0
3	1 min	9	-1
4	2 mins	8	-2
5	3 mins	8	-2
6	4 mins	7	-3
7	5 mins	6	-4
8	6 mins	5	-5
9	7 mins	5	-5
10	8 mins	4	-6
11	9 mins	3	-7
12	10 mins	2	-8
13	11 mins	1	-9
14	12 mins	0	-10

Figure 1.2 *Number Box* spreadsheet that records the migration of woodlice from light to dark conditions every minute.

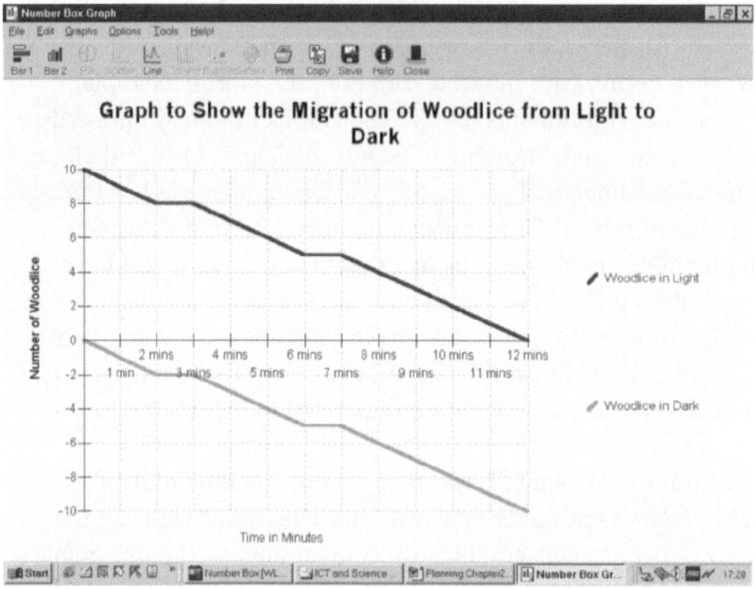

Figure 1.3 Line graph to illustrate the migration of woodlice from light to dark conditions.

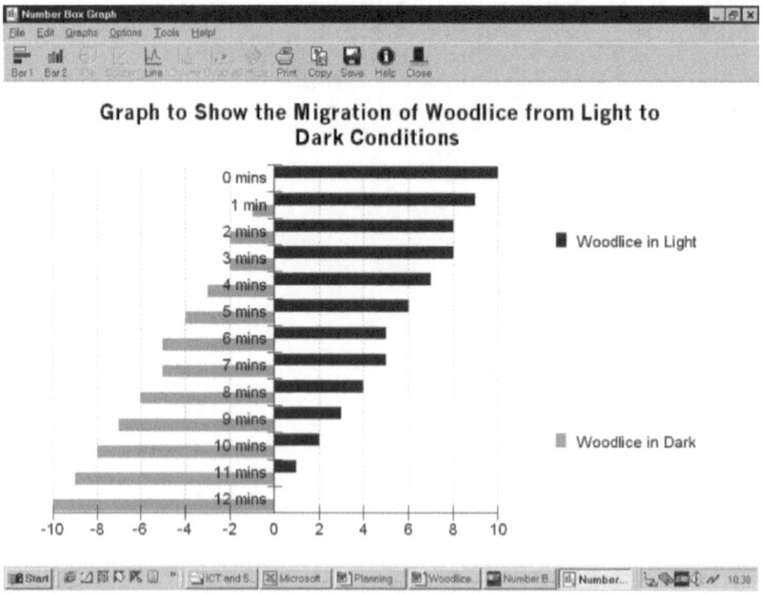

Figure 1.4 Column chart.

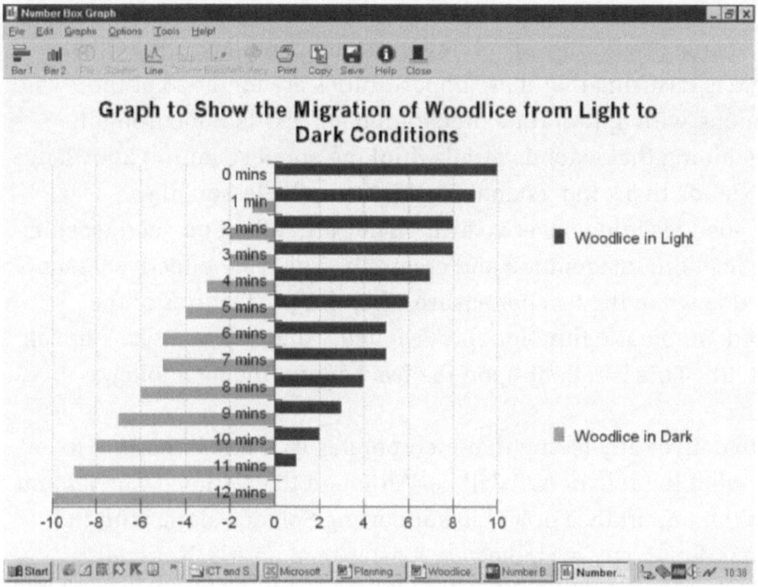

Figure 1.5 Bar chart.

Teachers may prefer to use the bar chart in preference to a line graph. The bar chart simply illustrates the numbers of individual animals, whilst the line graph might suggest that there is a mid-point between the minutes, when it seems that a part of a woodlice is present! However, what a line graph can be said to show is the progress of the migration from one environment to another. Whatever graph is used, the decision taken can form the basis of a very interesting discussion amongst the more able children in the class.

The very first sentence of this chapter made the point that ICT cannot and will not replace the teacher. Hopefully the example lesson plan above has demonstrated that the successful integration of ICT into science is relatively straightforward, and that much of it is reassuringly familiar, containing as it does all of the key features that one would expect in a plan for a primary science lesson. It has key learning objectives, details of prior learning, methodology, teaching points, foci for assessment, differentiation, follow-up work and resources. None of this is new. If anything, it is the relatively low profile of ICT within the lesson plan that should reinforce the notion that ICT needs to be embedded within the subject, not dominate it as though it was something that contains magical or mystical properties. The science provides the focus, content and detail, with the ICT providing the value added component, through its ability to organise, display and analyse the information that the children have collected. At all times the children remain engaged in the key scientific enquiry skills of hypothesising, predicting, observing, recording and analysing. The teacher's role here is ensuring that these opportunities are included at the planning stage, along with appropriate preparation and skilled, focused but open-ended questioning that extends pupils thinking about scientific knowledge and skills, and extends their understanding. Whatever the lesson, these fundamentals of good teaching never go away. They are reliant on good teaching, which means dedication, imagination and creativity. The truly effective lesson where ICT is used is when the teacher ensures that the key features of the computers – speed, automatic function, provisionality and interactivity – are all harnessed to the full. This is reliant upon the teacher, not the technology.

When deciding to deliver any lesson that incorporates ICT, it is important to be aware of exactly what the role of ICT will be. Although the National Curriculum clearly mentions the importance of ICT in supporting Science along with the other core and foundation subjects, the very existence of an ICT National Curriculum document ensures that it has the status of a subject in its own right. As far as this book is concerned, we are primarily interested in how ICT can support the teaching of primary school science. However, conversely there are elements of the ICT National Curriculum that can be successfully and effectively delivered through the use of primary science. Therefore, the teacher needs to fully understand how ICT is being used in the science lesson and that the teacher's organisation and planning both reflect this.

From the outset the teacher needs to be clear exactly how the computer is to be used, as well as why it is to be used. In the case study, it is clear that the ICT is to be used as a means of analysing and displaying information. The children observe the woodlice and then record their movements. Upon returning to the classroom, they have several key decisions to make, all of which need to have the teacher's input to ensure that appropriate and valid decisions are made. They will have to decide what type of application is to be used and why (in this case a spreadsheet as it is the only means of handling this kind of data and displaying it in a meaningful and relevant way), then they will have to prepare and then decide how to enter the raw data into the computer (they will have to use minus numbers to illustrate the progression from light to dark). When they have done this, they will have to analyse the data and interpret their findings – in other words, determine what the graph is actually showing them.

A computer is a powerful and expensive tool that should not be used simply because it is there. Employing a distinct 'value-added' component, it offers teaching and learning within the lesson, something that otherwise would not be available. In the case study above, for example, data collected as a result of practical science work can be sorted and displayed in a number of different ways at the touch of a button. Information might be displayed as a graph or as a chart, and several different versions of each can be produced, depending on how the user wishes to sort and display the information. Drawing graphs by hand can take a whole lesson, but by using a computer the time saved can be spent by the children engaging in higher order skills such as the analysis and interpretation of the information that is displayed in front of them. So, although the case study was taught effectively in the time before ICT was widely available in primary schools, it is the subsequent wide-scale availability of it that has enhanced the teaching of this topic. The practical science element has not been replaced, but the power of the computer has been harnessed to enable pupils to access levels of understanding that would not otherwise be available to them. It is the teacher's responsibility to ensure that this power is harnessed, and this is discussed below.

(1) Planning the science lesson where ICT is to be used

If ICT is to be used to support any given curriculum area it must support good practice in teaching and needs to be present at the planning stage, as is the case in our lesson plan above. Before planning, the teacher needs to be clear on how and why the lesson is being taught. It will usually form part of a sequence of lessons, rather than as a one-off, although it is possible that ICT may only be used to support the teaching content for a couple or even only one of these lessons. For example when studying the woodlice, the children observe and recorded their findings initially on paper before transferring the data into a spreadsheet for further interrogation. Although ICT is an integral part of the whole lesson it is only used at the final stage: in this case the preparation and entering of the data and the subsequent analysis and interpretation of findings. The teacher considered key features including how the lesson was organised, managed, delivered, assessed, resourced and followed-up. A more detailed discussion of these elements follows below.

(2) Objectives

The key objectives for a lesson or sequence of lessons will have been informed by long- and medium-term plans, which in turn will have been prepared subject to the requirements of the Science and ICT National Curricula. These determine what should actually be taught, and will come from a Scheme of Work, either written by the Science or ICT co-ordinator, or more usually drawn from those published by the Qualifications and Curriculum Authority (QCA). These will identify what needs to be taught, and the way that it could be taught. In this particular instance though, we are only interested in the objectives. In our lesson plan the science component has been identified and referenced to the relevant attainment targets from the National Curriculum Science document.

(3) Prior learning

Whatever happens, all lessons have to be informed by something. It is inappropriate for the teacher to decide that they are going to teach a particular topic, skill or concept just because they feel like it or because it is a particular strength or individual enthusiasm. The failure to use a medium-term plan and a scheme of work can lead to a lack of continuity and progression, where skills, knowledge and understanding is either omitted from year group to year group, or is repeated in different years without any progression on the part of the teacher or the learners. However, where they are used there should be a structured approach, so that concepts, skills, knowledge and understanding are consistently taught through introduction, development and extension. In our lesson plan the teacher considers the prior learning needed to ensure that the children learn what they need to learn to fulfil the required objectives.

(4) Development of lesson and (5) teaching points

These need to be considered together, as the former directly informs the latter. Point 4 is the content of the lesson and how it will develop, point 5 represents the 'nuts and bolts' of the lesson, that is what the teacher needs to say or do to ensure that learning takes place. Once the teacher had decided upon the content to be taught during the woodlice lesson, then its structure was planned. The objectives for the lesson were already known, but it was at this particular stage that the organisation and management of the lesson was planned. As can be seen from the lesson plan, there is a clear and logical progression to the lesson, conforming to pre-determined learning objectives, and the actual content is planned and prepared, including the key teaching points.

(6) Differentiation

The lesson plan makes it clear that there is little need in the first instance for differentiation. This is because the children can assimilate an activity such as this at a number of levels, but as the lesson plan quite clearly states, 'During the collection of the woodlice and the setting up of the experiments there should be little need for differentiation. However, for the reinforcement period, and the subsequent recording, specially designated worksheets and a CD-ROM will be made available.' As a direct consequence of this, the ICT that is to be used to support the science lesson, needs to be appropriate to the capabilities of the pupils and to the teaching and learning objectives for that particular lesson. However, when ICT is employed, particularly where it is used in a teaching and learning mode rather than purely as a tool, there is a strong danger that it will be incorrectly matched to ability. This can be due to the teacher's general lack of experience in the use of ICT within the lesson or because the teacher has a particular knowledge of a piece of software that they know will work, and thus tries to use it in the wrong context. An example of this might be in choosing to use a Key Stage 2 piece of software for Key Stage 1 or vice versa, even though the scientific concepts to be taught are relevant and appropriate.

It is important to think about differentiation at the planning stage. This is always difficult, as one has to consider not only the science but also the ICT content. In a typical primary school class, it is entirely possible that there may be pupils who are high attainers in science but lower attainers in ICT. Conversely, high-attaining ICT users may struggle with scientific concepts. This makes differentiation doubly difficult, as there are now two key elements of the lesson that have to be considered – the science and the ICT. The best solution to this problem is almost certainly to differentiate on the science rather than the ICT – if the science is the key focus of the lesson, then this is where the differentiation should take place. Additionally, in the case where the child may be stronger in ICT than science, the power of the computer can be engaged to develop the appropriate scientific knowledge, skills and understanding. This is where the skill and professional judgement of the teacher becomes crucial – and as mentioned above, is the reason why the teacher can never be replaced by a computer.

(7) Assessment

In order to differentiate effectively, it is necessary to ensure that there is an ongoing assessment strategy, as it is this that enables the teacher to prepare work for individual pupils that is appropriate to their ability. This is a vital part of the planning–delivery–assessment–planning cycle and needs to be considered at the planning stage. This is because assessment needs to be directly linked to the objectives for the lesson, as this determines how successful the lesson has been, as well as determining the levels of pupil attainment. For example, in the case of our lesson plan, it can be seen that the four objectives of the lesson are:

- To understand what kind of animal it is (the woodlouse)
- To understand something about its natural environment

- To learn how to undertake a 'fair test' in order to discover something of its preferred habitat
- To learn to take care of, and have consideration for the well being of all types of animal.

And the assessment criteria are:
- Can the children appreciate the concept of a fair test?
- Why is a fair test necessary in this instance?
- Why have the woodlice migrated?
- What are the graphs showing?

This overtly connects the objectives with the outcomes and poses several key questions:
- Have the children learned what you wanted them to learn?
- Has your teaching been effective?
- Did you cover all of the intended learning outcomes?
- At what level are the children working?
- What additional practice may they need?
- Which concepts, knowledge and skills need to be reinforced? Which ones need to be extended?

If the teacher is to plan effectively for children of differing ability levels then they clearly need to know the level at which the children are working. If a child is completing a task that is inappropriate because it is too easy or difficult, then learning will not take place. However, where ICT is used the task of assessment becomes especially difficult as there can often be some confusion as to whether it is the subject component, in this case science, or the ICT element that is being assessed. As Crompton and Mann (1997) have stated, the problem is that many teachers see the word 'technology' and misinterpret exactly what this means. This of course will necessarily depend on the focus of the lesson's objectives, as the purpose of assessment within the lesson context will be to discover if learning has taken place. This formative assessment will in turn inform future planning. When assessing science it may be pertinent to mention that a child's progress may have been hampered by, for example, an inability to enter information due to unfamiliarity with the keyboard or the program. However, as the focus remains on the subject (in this case science), the assessment criteria should be made explicit in the planning stage. The teacher will determine what other criteria need to be used to indicate whether or not progress has in fact been made.

The assessment of children's work when using ICT can take several different forms: formative assessment as described above, summative assessment, where some form of task or testing takes place in order to group levels of achievement into broad bands, such as National Curriculum attainment targets, and diagnostic testing to identify the cause of a particular problem. These will usually be directly linked to planning, and will involve observation and/or questioning, or perhaps self-assessment by the pupils themselves. As with all investigative activities in science, it will be necessary to determine key aspects such as which (or if any) problem solving techniques were used, how long the children were on task and the levels of collaborative discussion of the pupils. Whatever form the assessment takes, it should evaluate the activity in terms of its relative success, inform the planning of future activities and identify those individuals who might need extra help, i.e. reinforcement activities. The focus for assessment was considered, as well as how the content would be differentiated. The key content of the lesson was delivered during an input phase, where the teacher put across to the children the key teaching and learning points. The children then completed the task, in this case the investigation itself, and then entered the data and subsequently interrogated this data on the computer. The lesson then concluded with a plenary to discuss what was learned. However, during the course of the lesson some form of assessment should also have taken place. As a result of this assessment, the teacher will then have the information to plan and deliver the next lesson in the sequence, and will have identified any of the class who may need extra support and reinforcement activities. No doubt there will be others who will be ready to progress to more complex tasks. The teacher is then ready to re-enter the cycle of planning and delivery, hopefully at a higher level, for if it is at the same level, then it is likely that little or no learning will have taken place.

(8) Cross-curricular links

It is important to identify cross-curricular links at the planning stage for several reasons. Children do not learn isolated skills or parcels of knowledge in a manner which is divorced from either other subjects or the world around them. Daily exposure to media such as television, video and the Internet ensures that learning occurs in a cross-curricular way. Likewise, our lesson, although primarily scientific in focus, contains elements of ICT (as explained), Mathematics (negative numbers, graphs) as well as Geography (light and dark, warm and cold, micro and macro climates) and Citizenship and PSHE (working together, consideration for other living things). It also enables the teacher to cover other areas of the National Curriculum at the same time, and the skilled teacher can plan for this, thus saving time and energy! There are many such examples listed in the ICT National Curriculum document.

(9) Resources

The lesson also needs to be resourced, and this should be made explicit on the lesson plan, which then should be remembered and read! Although a cliché, the phrase 'fail to prepare, prepare to fail' has a strong element of fact about it, particularly when a practical element is involved. The teacher must consider carefully what resources are required, and these must be prepared and readied prior to the start of the lesson. Failure to do so can lead to the teacher feeling at best under-prepared and thus lacking in confidence, but at worst can mean that key equipment needed for the lesson is missing, preventing effective and efficient teaching altogether.

As far as science and ICT are concerned, this could be quite far-reaching and detailed. For the science component this might involve the use of scientific apparatus, investigative equipment, research and other reference sources such as books, charts or posters. Fieldwork equipment such as clipboards, worksheets, pencils and paper, may also be needed. For the ICT component it will obviously involve computers and software, but might also involve other resources, depending on the type of science lesson that is being taught. These might include a digital camera, a scanner or printer or involve CD-ROMs, data-logging sensors or computer microscopes (an electronic microscope that can be directly connected to a computer, with the image being displayed on the screen with the facility to zoom and record whatever is underneath the lens). This is discussed further in Chapter 9. There may be other less obvious resources required, such as clip-art or information from the Internet. Whether they are all used, or a selection of them or only one or two, then they all have to be considered at the planning stage and access to them needs to be ready prior to the start of the lesson. The lesson will not be able to proceed as planned if the resources are not in place and this will fundamentally affect how the lesson will be delivered. After all, if computers are not switched on or logged on, or printers do not have any ink in them, or the batteries for the digital camera are flat, then this will prevent the intended teaching objectives from taking place and thus will delay the progress of the lesson. This could be very significant if the lesson is taking place in a heavily timetabled ICT suite. It is of course important to ensure that the resources that are selected are appropriate to the lesson's objectives. For example the software has to be relevant – if a report is being produced then clearly some kind of word processing, desktop publishing package or presentation software will be needed. This may seem obvious, but where some scientific concepts are concerned, for example data handling, specific decisions pertinent to this may be needed. What is the most appropriate – a branching database, a random access database or a spreadsheet? This needs to be ascertained from the outset, and the only way that this can be properly done is for the teacher to practise in advance so as to determine which software will be the most appropriate for a given task. This will be discussed further in the next chapter.

Organising and managing the lesson

The discussion above describes and explains how ICT can be used to support a science lesson, but there are also other aspects that need to be considered. Although the learning objectives for the lesson will be scientific, ICT has to be included in the lesson plan in order to indicate its role. It is not appropriate to have ICT as an 'add-on' in the sense that it is an afterthought. It is not only a tool to the successful teaching and learning of science, but it can actually influence and enhance the way that science is taught and learnt. This is a crucial point, for although it can be employed purely in the role of a resource, as we have already seen, it offers both teacher and learner so much more. Therefore the teacher needs to demonstrate that the place and role of ICT has been clearly and deliberately thought through, and this needs to be clearly indicated at the planning stage. It will not only be *ICT – how?* but as was discussed in the Introduction, *ICT – why?* The teacher will be that much more effective if this role has been clearly indicated, in that the teaching will subsequently have a better subject content and delivery. As with all lessons, thorough preparation and planning ensures that the teacher will have a safe and secure subject knowledge, and the lesson will be better for it. The teacher will appear confident and knowledgeable, the content will be relevant, the lesson will be well paced and effective learning will occur.

We have already seen that good primary science is a practical, hands-on subject that employs the key skills of hypothesising, predicting, analysing, observing, recording and drawing conclusions that are valid, reliable and meaningful. As a result of this, there need to be opportunities for collaborative learning to ensure that children can work together on a task and discuss their work amongst themselves. It is important to stress at this point that genuine collaboration is an educational device, not an organisational one. Along with the other subjects of the primary curriculum, the most meaningful science occurs when children work together on a shared task and can discuss it together. It is not just a question of organising the class into groups, where children are nevertheless working individually, but of ensuring that the children are actually working together. When ICT is used this becomes especially important, as interaction becomes the key. Science should not be an isolated activity, and when ICT is used there is the additional danger that this may take place. Children should not become isolated and passive recipients of the material on the screen; they should interact with the computer as much as with their peers, by using it in a context that demands composition, interrogation, analysis and the drawing of meaningful conclusions, rather than requiring simple yes or no answers or the memorisation of simple, isolated scientific facts.

Related to this interactivity is the need for the teacher to be clear about which key questions are to be asked to promote and stimulate learning. This key aspect is frequently overlooked. A focused, yet open-ended question, whilst the task is being undertaken, can enable the children to extend their thinking and subsequently their learning. For example, a teacher talking to children working at a computer can ask questions such as ...

- 'That looks interesting, tell me about it'
- 'Why do you think that you got those results?'
- 'What would happen if ...?'
- 'What might happen next? Why?'

The next key issue that needs to be considered is how the ICT that is being used is being organised and managed within the structure of the school. This will often be determined by where the lesson is actually taking place. In the primary school this will involve a dedicated ICT suite or the classroom, or even perhaps another part of the school such as a corridor or the hall. Much of the science may be carried out in the school grounds if work with minibeasts or plants are being covered, in much the same way as our woodlice investigation. If a school has a set of laptops, possibly with wireless networking, then work can be saved immediately to a central file server. Instantly, children can extend their studies by using this wireless networking facility to undertake relevant research from the Internet, perhaps by visiting relevant web sites or emailing an expert at a museum or university. This opens up a whole series of new and genuinely innovative opportunities for learning, which will depend on where the science actually is. The scope is enormous. None of this will restrict the use of ICT, but it will have to be taken into consideration when planning the project. The kind of science that is being taught will often determine the way that the lesson is organised and managed. Whatever the lesson, how it is organised must reflect the learning that is to take place.

This can take several different forms.

- Where the children's main role will be as learners. This is where the teacher demonstrates a skill and the children subsequently acquire and reinforce this skill. This is especially useful where the teacher will be introducing a new science topic and/or a new piece of software with which to develop their investigations.
- Where the children are involved as peer tutors, where a group of pupils learn the key skills and then 'tutor' others. This is particularly effective where there may only be one or two computers in the classroom.
- The most popular organisation will be in pairs, which allows for pupil interaction, discussion and shared activity. Couples can be arranged according to their similar abilities, or be of mixed abilities, or by peer tutoring either in their capability at science or ICT. Although it is possible to arrange the children in groups of three or more, this should be largely avoided during the ICT element, as some adult supervision will be needed to ensure equal access to the keyboard, mouse, and the information that is being used for entry into the computer.
- Occasionally, it will be necessary or desirable for individuals to work alone so that individual ideas can be developed or reworked.

However the learning is organised, the teacher needs to ensure that it is suitable for the age and abilities of the children, and that an appropriate teaching style is being employed. Admittedly this can be difficult; there may be a lack of resources and time, and the form that the lesson takes may be restricted by where it takes place. If teaching is to occur in an ICT suite then groupings may be necessarily restricted by the layout of the room.

Organising learning with a classroom computer

In an ideal world the most desirable way of using ICT may be within the classroom itself, or even in the wider environment as described above. This will ensure that ICT will be used within the context of the science being taught as well as being in the same location as the investigation being followed. It will be integrated with other classroom activities and it will usually be near other resources so that research and all aspects of collaborative learning can easily occur. However, there are inherent problems with this. Assuming that there may only be one computer in the classroom, there will be a key issue of time and access to the technology. When time is taken out from the school day for assembly, PE or games and other timetable 'blocks', i.e. literacy or numeracy (it is the authors firm belief that science and ICT can make significant contributions to both, but more of that elsewhere), the use of what little time that remains becomes crucial. As a result of this, the teacher needs to maximise the 'hands-on' time as much as possible. In order to do this, there are several strategies that the teacher can employ. The following scenario will illustrate this point.

During a science project based on National Curriculum (Sc2), a class teacher wishes to teach her Year 3 class how to use a program that simulates the construction of a skeleton. The task involves children pointing and clicking on a bone then dragging and dropping it onto an outline. By doing this with all of the key bones of the body, a two-dimensional skeleton can be constructed. These key bones can then be labelled and either saved or printed for the children's files as a record of their work. In order to do this, the teacher gathers the class around the computer and then demonstrates the key features, such as how to load the software and how to drag and drop. Older children can be asked to take notes that they can then use later. At this point the teacher may either wish to introduce another activity or send the class to begin work. Whichever way the teacher chooses, what happens next is crucial to the successful management of the class. The teacher will want to maximise the time on the computer, so they will send two children to begin work, while dealing with the other thirty or so children in the class.

It is important at this stage that the teacher does not become involved with problems concerning the computer whilst having to deal with thirty other children – this is not an effective use of the teacher's time and could lead to disruption at this crucial phase of the session. Therefore the teacher needs to send two of the more able children in the class to begin work on the computer. One will complete the task and will 'drive' the mouse, whilst the other one helps by contributing scientific data or knowledge and ideas. In this way, one child does not become an isolated user at the screen, for at the same time the other is able to learn about the workings of the program. When the child 'driving' completes their work, they save or print it and he or she then move aside. The second child moves into the 'hot seat' and another child comes to act as helper. It is important to change this 'elite' group from time to time, but the fact that more able children are being used at this stage will ensure that they will be able to sort out any problems that arise. This will ensure that the teacher can concentrate on getting the remainder of the class settled and on task, before moving across to see how the children who are using the computer are getting on.

By this time the pupil–teacher interaction should be concerned with talking about the task, and not concerned with technical aspects of the ICT, such as how the program actually works. This is a vital part of the lesson. Not only can the quality of the pupil–teacher interaction be that much better, but it also ensures that the teacher gets that vital two minutes or so to manage the transition from sitting, looking and listening to the demonstration, to the children actually starting work. Those first couple of minutes are crucial to the success of the ensuing lesson.

Organising learning in an ICT suite

Different issues arise when using an ICT suite. First and foremost, it is a completely different environment. As with other lessons taught away from the classroom, there are different organisational and management matters to consider. In that sense, teaching in an ICT suite is more akin to a PE, art or music lesson in that different teaching and learning strategies are required. This is not necessarily difficult, but it is different, therefore the organisation and management needs to be carefully considered at the planning stage. As previously discussed, both the science and ICT component need to be considered together, but in this particular context the computer becomes the focus of all activities and it can be difficult for both teacher and pupil to appreciate that this is a science lesson.

For a start there are issues that the teacher can do little about. Unlike their classroom, children are restricted to where they can sit by the position of the computers, so they may be facing a computer screen rather than the teacher. Therefore the teacher may need to ensure that the pupils turn to face the teacher for any direct teaching and/or whole-class input. This may sound obvious, but depending on the layout of the room there may be corners to hide in or computers to hide behind.

Another organisational and management issue is that there may only be a sufficient number of computers for half of the class, which may entail splitting the class into two groups. Whilst this certainly makes the use of ICT that much more effective, it leads to other organisational issues, such as the key question of who teaches the other half of the class? This might entail the class working in pairs, but again this might be difficult if the computers are too close together to allow this. If there is an element of practical scientific investigation involved in the lesson, is there any desktop or bench space to set up and carry out scientific investigations? There will certainly be this space in the classroom, but it is highly unlikely that it will be available in the ICT suite, especially as in many cases it will be a room converted from a previous usage and will be too small.

There will also be technical issues to consider, such as who turns the computers on, changes ink cartridges, manages the network, troubleshoots problems and generally looks after the facility.

Conclusion

Wherever the lesson is delivered, it must be remembered that the science element always comes first. It is not an ICT lesson, for although it may offer opportunities for ICT key skills to be developed or the achievement of National Curriculum attainment levels, the key objectives of the lesson should always concern science. We have discussed at length above the contribution that ICT can make to teaching and learning in general, and science in particular, so it is therefore essential that this power is harnessed in a subject context. The need to plan and deliver the science lesson carefully never goes away – a scheme of work needs to be followed, a medium-term plan prepared, a lesson plan written and delivered. However, the use of ICT can enhance the teaching and learning process greatly.

Bibliography

ASE, *Be Safe!* third edition, Association for Science Education, Hatfield, 2001.
Crompton, R. and Mann, P. (eds), *IT Across the Primary Curriculum*, Cassell, 1997.
Department for Education and Science, *The National Curriculum for England*, DfES/QCA, London, 1999.
ECB/Channel 4, *Howzat! Playing the Game*, ECB/Channel 4, Warwick, 2002.

Chapter 2

Selecting software for science

Once the decision has been taken to use ICT as part of the science lesson, the next step is to decide which software is to be used. This needs to be considered carefully, as the software is the medium by which a significant element of the lesson will be delivered. Mention was briefly made during the previous chapter of the need to select and use software that is appropriate in ensuring that the teaching and learning objectives are met. This chapter expands on this.

The importance of choosing appropriate software should not be underestimated as this acts as the interface between the user and the computer. Apart from detailed planning, preparation and teaching, it is crucial that appropriate resources are employed as an integral part of the learning process. We particularly want to suggest which software to use for science education and how teachers should go about selecting suitable programs. We have already mentioned that the computer has the power to handle huge amounts of a wide range of data types quickly, automatically and interactively. It is the range of software such as word processors, desktop publishing packages, spreadsheets, graphics packages, email and Internet browsers as well as hardware such as digital cameras, scanners and electronic microscopes and their associated software that enable the user to take advantage of these principal features, and all of these can be used as part of science education. In turn, the user interacts with the software to bring creativity, original thinking, and raw data to the learning situation. The computer will provide the power to sort, search, interrogate, analyse and display this data as refined information in a variety of forms. The creativity and ideas that the user possesses will contribute to the choice of the software that will allow the computer to function interactively. The quality of the lesson, and in particular the quality of the ICT component of the lesson, will stand or fall by the standard of the software and how it is used. Relevance to the science being taught, and whether it is appropriate to the age and ability of the children, must all be considered when choosing software. This is why it is very important for teachers to ensure that they select software that is appropriate to the task that is being set, and to the objectives of the lesson. It must have the necessary flexibility and be sufficiently interactive to make it is easy enough for the children to be able to interact with it effectively. The better the software, the better the lesson, the better the learning.

Historical context of software for primary science

Strangely, even in the twenty-first century and as part of an education system that has been using computers for over twenty years, it is only relatively recently that serious attention has been paid to the importance of educational software. In the early days of ICT it seemed that little regard was given to the content of the programs, with a lack of theoretical background and certainly no understanding of their application to the needs of primary education. Early science programs were often simply crude pictures together with a question or two that had little interaction with the viewer. At best these could be used as a simple reinforcement of previous work, but at worst might be used instead of the actual practical science itself. These programs were often written by specialist computer programmers who had little or no background or experience of education, and there was correspondingly little or no teacher input into the process. This was just one of several key reasons why the computer was often not well used in those early days. At this time there was often a belief pervading that the use of computers in schools was a good thing, regardless of the actual educational content, application or relevance. Indeed, during the 1980s many teachers were taught and encouraged to write computer programs in BASIC for the BBC computer, the idea being that they would then be in a position to amend or even to write their own programs. This approach set ICT back, as understandably, few teachers had the time or the inclination to do this, and many were put off by the apparent need to be good at computer programming in order to use ICT effectively with their children. This, when coupled with a lack of critical analysis of software, led to some very poor ICT being taught in schools.

Nevertheless, computing had to start somewhere, and there were a few schools who, realising that programming was inappropriate, did the best they could with the few programs that were available. These were often simulation exercises, such as the one based on the 'Mary Rose'. Although far from perfect, it was one of the first programs of its kind specifically designed for schools, and contained a considerable amount of good-quality background material, which was quite innovative at the time. Apart from this, these schools often pioneered the use of the computer as a word processor. Perhaps for the first time, children could be observed working in small groups at the class computer, collaborating in producing a story, a book, or even a school newspaper.

However, from the late 1980s onwards, software was delivered or 'bundled' with new computers, a practice that largely continues to this day. Initially many Local Education Authority ICT Advisers often selected a range of programs in consultation with the software and hardware companies. This had several advantages as it meant that experienced and knowledgeable ICT teachers could evaluate software in depth and could then make recommendations as to which pieces of software could be included in the bundle. They could take the key decisions of which software was or was not good, and which programs were appropriate for different contexts without the class teachers having to plough through lots of unfamiliar software until they found a program that was useful. These advisers were then in a position to train classroom teachers how to use ICT effectively with their classes. However, a drawback of this approach was that any evaluations of the software could only be of a very generic nature, and as a result did not have any subject-specific application or emphasis. Although this fitted in very well with the topic-based approach that was prevalent in primary schools at this time, it lacked any kind of specificity and subject focus. If the teacher was confident in the subject this was seldom a problem, but if they lacked knowledge of a specific subject there could be difficulties. It was often science that suffered in this way.

More recently, computer hardware companies have made this selection, and although this has often been in liaison with advisory staff it has only occasionally involved the purchasing schools. Whoever has taken these key decisions, the fact remains that this process has ensured that there has often been little need for teachers to engage in any form of critical analysis of software. The selection of the software, although probably based on very sound educational considerations, gives no account to the local needs and requirements of individual schools. The inclusion of a particular program in the bundle might be determined by economic or political rather than educational considerations. The assumption has often been that as the software packages have been chosen on their behalf by either LEA advisory staff or hardware companies, then they must be good. However, those teachers that have to use the software on a day-to-day basis are being denied the opportunity to apply any form of critical analysis to the software selection process and subsequently extend their own understanding of software evaluation. It is unlikely that this would happen in other areas of the curriculum. Would primary teachers allow themselves to be denied any input into the selection of a Mathematics or Language scheme? Highly unlikely!

Up until the advent of the Windows operating system, and in particular the widespread adoption within society of *Microsoft Office*, the software that was available was mostly written specifically for children. In recent years though, the software employed in schools has moved closer to the industry standard that is used every day in the 'real world' such as *Word, Excel* and *Publisher*, and employs common tools such as the Internet and email. A particularly good example of this is the *Black Cat Toolbox 2000*, currently published by Granada Learning. The word processing program as part of this package called *Write Away!* closely replicates *Microsoft Word*. However, as far as science teaching is concerned, the spreadsheet that is included as part of this package, *Number Box,* is quite closely based on a 'grown up' spreadsheet such as *Excel*. This includes several excellent features which makes it very easy for even young children to use, such as 'Quick Sheets'. These are pre-prepared spreadsheets that are attractively presented and have cells that only require filling in before pre-programmed formulae calculate the answers. It is this kind of software that ensures that the power of the computer that is available to commerce and industry is now also available to even the youngest primary school child and their teacher. An added advantage of this is that the children will be using software with a similar layout and approach to that which will follow them throughout their school career and beyond.

What do teachers need to consider?

So when choosing software to support the science lesson, what should the teacher consider? For, according to the BECTa web site (2002):

> For teachers, the selection of software for classroom use may have an unwanted element of chance. Choices which are based on information provided by advertising, publicity material or packaging can turn out to be inappropriate, which is both time consuming and costly. Such descriptions may not adequately reflect the potential educational effectiveness of software in use. Choices may be made in circumstances where there is little time available to discuss aspects of the software in use, or to try it out.

It goes on to say,

> Some of the most useful software evaluations for teachers are those that provide an indication of the context for software use in the classroom. Teachers organise classroom conditions to ensure that their pupils interact productively with software; sharing an understanding of these specific

contexts for software use can help other teachers in their selection of appropriate software. Teachers therefore require:

1. information about what software is available
2. information about use of software in educational contexts.

http://www.becta.org.uk/technology/software/curriculum/evaluation1.html (accessed May 2002)

BECTa also say:

> Before defining the minimum requirements for software to deliver the National Curriculum, it is important to be clear about why we wish to use ICT in science and when it is, or is not, appropriate. It is only appropriate to use ICT:
>
> - when there are clear gains to teaching or learning
> - when it enables the teacher to improve their teaching and/or enhance pupils' learning.
>
> Any software should take into account the above points. The ICT itself should not be the focus of activity but should be a transparent platform for solving problems or communicating ideas – activities which are embedded in all classroom work.

It goes on to say:

> Owing to the many opportunities for using ICT in science, a minimum 'toolkit' would include most applications. This should consist of content-free, generic software which can be used in many different ways for a multitude of tasks. The software should meet three general requirements:
>
> - It should reflect the current industry standard
> - It should have a transparent and intuitive graphical user interface
> - It should have the capacity to offer a series of configurable 'front ends', to make it appropriate for pupils of different ages.
>
> The following software and hardware is required to support communication:
>
> - Word processor and/or desktop publishing
> - Presentation package
> - Painting/drawing package
> - Internet and email
> - Image manipulation (digital camera/scanner)

- Modelling package
- Multimedia (including CD-ROM)
- Sound (sound samples/music/speech).

http://bbwww.becta.org.uk/technology/software/curriculum/reports/science.html (accessed June 2002)

So what makes good educational software?

There are several key factors here that need to be considered. Clearly, the software needs to be appropriate to the objectives of the lesson and any tasks that will be given to the children. We make no apology for mentioning this yet again but this is fundamentally important. It should not only match the key functions as detailed above, but should match them well. It will need to support the teacher's requirements and so will be required to fit into the teacher's preferred teaching style and use an appropriate model of learning. So, if a teacher mostly employs a teaching style that encourages group work and collaborative learning then they will need to set up situations where the children can learn collaboratively, and the software that is used will need to be able to support this.

Conversely, the teacher may feel the need or wish to adapt their own style to fit in to the requirements and the structure of the program. This is an example of where ICT can develop staff and actively develop the teaching and learning of a particular teacher.

What about technical considerations?

When choosing software, the teacher needs to be aware of several key technical considerations. These are:

- Compatibility
- Network capability
- Processing speeds
- Graphics cards
- Sound quality.

The software needs to be compatible with the type of computer that is being used, a fairly obvious point but one that can be easily overlooked. Many CD-ROMs often come as dual-platform programs, in that they can be run on both Microsoft Windows and Apple Macintosh compatible computers. If an iMac is being used, the software needs to be in the form of a CD-ROM as iMacs do not have floppy disk drives as standard. Other considerations might include whether or not the program can run over a network, and if it can, whether it does so smoothly and quickly. It is very frustrating for a child to have a program that runs slowly, especially if it is a data handling package where information needs to be processed quickly and in a number of different ways. This is an increasingly important consideration in the primary school as many now have their own networks with the software being installed on a network file server. For example, some CD-ROMs cannot run over a network, and the CD-ROM may need to be loaded in the computer to run, whereas others need to have the screen resolution of the computer changed in order for the images to appear in the correct colours, or even run at all.

Many modern programs, especially CD-ROMs or web sites, incorporate high-quality graphics or images. These have technical implications for the computers on which they are being used. These often include video clips which need to run smoothly and in the correct time-frame (without being jerky or pausing whilst 'buffering' takes place – that is, loading the images into the computer's memory), so will need good-quality graphics and video cards installed in the computer. In order for these video clips to run, the computer requires video playing software such as *Quick Time*. Although freely available as a download from the World Wide Web, or included on the CD-ROM, it means that someone has got to find and install this software onto the computers. If the school has a significant number of computers, and many primary schools today have upwards of thirty or so, then this will mean a lot of work for someone! However, this is a worthwhile use of time, as once installed for one program, it can be used for several others and enables the user to achieve the fullest potential from the computer and the software. Although less of a problem these days a program may have graphics that need a special type of graphics card installed within a computer, and this may not come as standard. Additionally, if the program requires sound, then the computers will need sound cards installed to play any audio, for, as with the graphics and images, sound and picture quality needs to be very good otherwise there is little point in using it. There is no point in electing to use ICT if the quality of the hardware and software impedes the message that the teacher wishes to transmit to the pupils. This in turn leads to other related issues: sound means noise, which in turn means disturbance. The teacher needs to decide whether this is acceptable and needs to ask several key questions. Is the sound absolutely necessary? Can it be turned down or even off without lessening the effectiveness of the program? Are headphones needed? If so, this is another purchasing requirement, and of course only one set of headphones can be used per computer – not much use if any kind of collaborative or group work is the aim.

What about educational considerations?

These are the most important considerations of all, and are detailed below.

- Does the software teach the children what the teacher wants them to learn? Does it enhance the learning experience?
- Is the software easy to use?
- Is it intuitive?
- Does it use common commands?
- Is the manual short?
- Is there online or on-screen help, and if so, is it short, helpful and easy to read?
- Is the screen interface clear and easy to follow? Is it bright, attractive and appealing to the user?
- Can icons be clicked easily? Are they large enough for younger users to be able to use?
- Is the reading level appropriate to the reading age of the user?
- Is the content accurate? Is it free of bias?
- Is there an option facility so that the teacher can change the ability level? Can the teacher change the difficulty level of the language, or even the language itself if English is not the pupils' mother tongue?
- Can I get a trial copy of the software either as a download from a company web site or on approval through the post?

Firstly and above all else, the software must help to teach the children what the teacher wants them to learn. The end product from its use must match the intended learning outcomes that were detailed during the planning stage. How this occurs will depend on the objectives of the lesson. For example, if the teacher wants the children to enter data into a database, interrogate it and analyse the resulting information, then the teacher should select software that allows this to happen. They should also select a package that is suitable to the age and ability of those that are to use it. There is little point in getting, for example, a group of 6-year-olds an office standard database, when a product such as *Number Box* from Black Cat, which is described below, may be highly suitable. Alternatively, if children are trying to find factual information on a scientific concept or idea, or a piece of science history, then the resource that is used, be it a CD-ROM, web site or whatever, needs to give them the information that they need, be easy to navigate through and have text that is of an appropriate reading age. Additionally, it should not just *be* the learning experience, it should *enhance* it. It should offer both the teacher and the learner something that otherwise would simply not be available. This might involve multimedia in the form of sound and video, or the ability to search, sort, interrogate and analyse information. Above all there needs to be interactivity, where the child is not passively receiving information from the screen. This means using the computer not only as a tool for learning but to use it so that it significantly influences the teaching and learning process.

There are other factors that determine the quality of software. It must be easy to use in that it needs to be intuitive, that is simple for pupils and teachers to use with a minimum of instruction and explanation. One of the main purposes for using ICT is to enhance the teaching and learning experience, not to hinder it, so it is important that the computer does not act as a barrier to learning. Again, according to BECTa (2002), "interactivity should be meaningful, and should contribute to learning rather than adding distraction". This can happen if the chosen software is too difficult to use, or requires so much explanation that it detracts from the main focus of the activity. Most modern software uses common commands and functions for operations such as saving, printing, loading, running and copying and pasting and programs can usually be integrated with one another.

The user manuals and online help should be short, contain plenty of illustrations and preferably be written in a way that is easy for the children to follow. If the manual is large, it is highly unlikely that the teacher will have sufficient time in an already overcrowded day to read it, let alone familiarise themselves with the key features to such an extent that they will be able to go away and use it with children. Besides, if the program needs a user guide of this size then it will be too difficult to use anyway!

The screen should be uncluttered and easy to follow, with clear icons. If the program is intended to be used with younger children, then these icons need to be large enough for them to be able to point at and click easily without them becoming frustrated. The textual content of the software should be appropriate to the reading age and abilities of the children, as there is clearly little point in giving the children software to use that is too easy or too difficult to read. Additionally, the teacher needs to be aware of the factual content of any text or web site. They need to ensure that the material that is being used is accurate and reliable, and that it does not contain any bias towards or away from a particular viewpoint or idea.

There should also be an options facility so that the teacher can set the difficulty level appropriate to the age and ability of the class, or even to individual children. It may be appropriate to have the facility to change the language if English is not the first language of the pupils. Another point to consider is that many CD-ROMs and web sites have their origin outside of the United Kingdom and as a result contain language in American or non-standard English. Although these may still be suitable for use, the teacher needs to ensure that pupils are made aware of any differences between the two.

Before committing scarce financial resources to the purchase of software, the teacher should always take the opportunity to evaluate the package using the criteria listed above, in exactly the same way that they would with any other educational resource. Most software publishers make their products available on approval, either as samples through downloads from their web sites or through the post for a trial period, usually for thirty days. However, often these are not the full versions of the programs, or they are 'timed out' in that they stop working once the end of the trial period has been reached. Many of the best publishers' web sites can be found at www.teacherxpress.com (note the spelling!), by moving to the section called 'Educational Software'. Users can then access the required publisher and follow the instructions that are provided.

So which pieces of software can I use to support primary science?

Two excellent examples are *Number Box*, a spreadsheet package specially designed for use in primary schools by Black Cat software, and the *2Simple Video Toolbox*. Examples of these are given below.

Number Box Yellow level (Figure 2.1) is for early years children. There are very few, large icons and only two drop-down menus. *Number Box* Red level (Figure 2.2) is for use with older children. It has many more, smaller icons and correspondingly more options, as well as many more drop-down menus.

The *2Simple Video Toolbox* is powerful, easy to use software that is appealing to the younger user. It has large icons and a minimum of reading. This is especially important, as young children may not have fine control of a mouse, which is a piece of equipment that is designed to be used by an adult. It also provides the child with a range of other features that can be accessed simply by pointing and clicking. These include a range of templates so that a range of simple desktop publishing pages can be produced, or different stamps for pictograms when displaying any kind of data handling information. This is illustrated in Figure 2.3. *2Count* is a simple recording program, in which the children can select their own pictures in their pictograms. Eye colour has been selected for the graph, with other options that are relevant to science being displayed in the selection

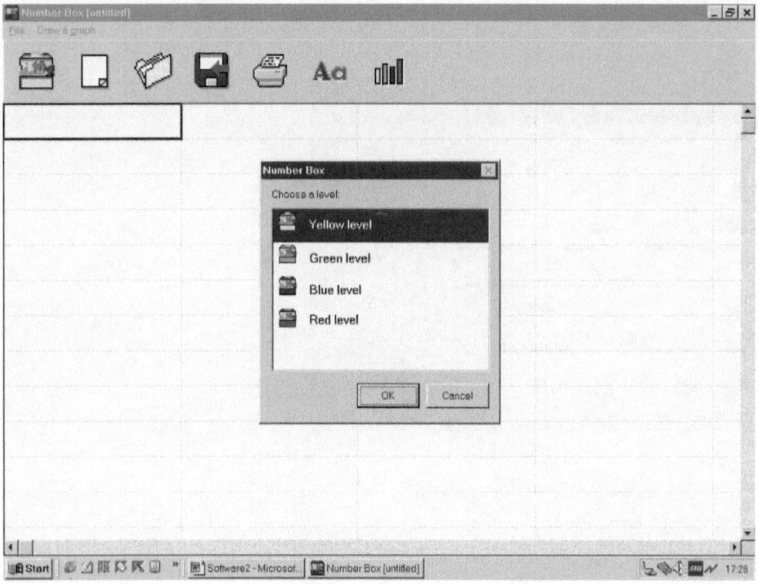

Figure 2.1 *Number Box*: Yellow level.

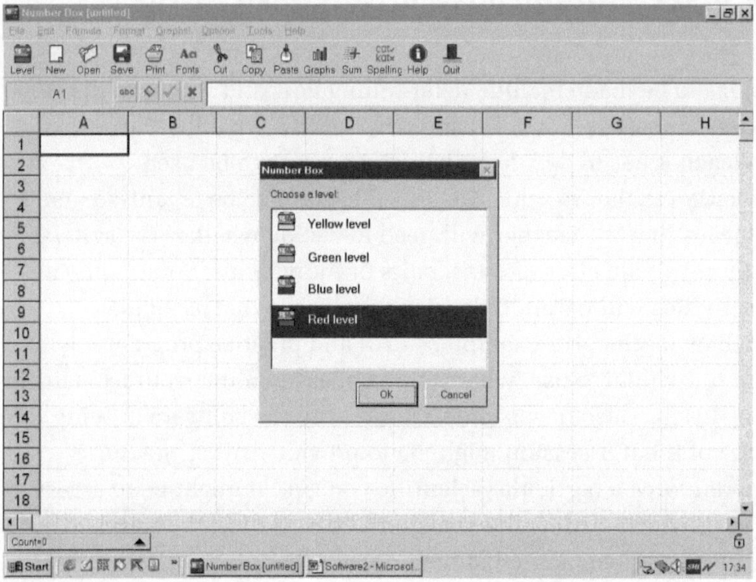

Figure 2.2 *Number Box*: Red level.

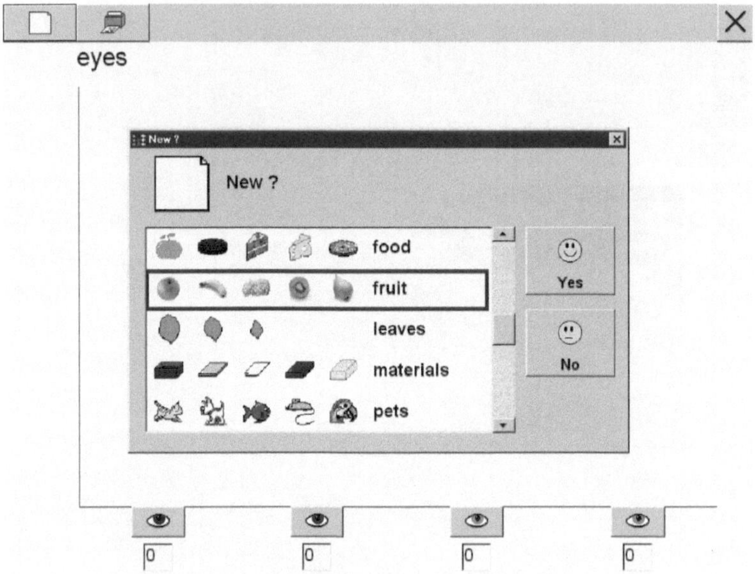

Figure 2.3 2Simple Video Toolbox: 2Count.

window. Other parts of the program include a simple branching database, graph builder, paint package and a LOGO program.

What is the place and purpose of software in the curriculum?

Other key decisions to be made include determining how and where in the science curriculum the software will fit, and particularly, what the purpose of use will be. The teacher has to decide how the software is to be used, particularly in relation to the way that the lesson is going to be organised and managed. This means that the teacher will need to decide whether the software is to be used with the whole class, groups, pairs or individuals. Some software and associated activities will naturally lend themselves to one particular approach rather than another. For example, a drill and practice program might be suitable for an individual to use, whereas a database might encourage pairs or small groups. The teacher also needs to determine why the software is being used. The purpose of the ICT session might be to provide drill or practice opportunities, to reinforce a particular scientific concept. It might be to assess that concept or to provide a variety of extension opportunities. Whatever the context, the teacher needs to ensure that both organisationally and educationally, due regard is given to the type of software that is employed and in particular, the way that it is used.

In 1977 Kemmis, Atkin and Wright devised a classification of educational software. Although old by the standards of modern educational ICT, it is still quoted in a range of educational texts and provides a useful framework by which to categorise different types of educational software and by definition, different types of teaching and learning situations. Four categories were devised in what is essentially a continuum, with instructional software at one extreme and open-ended software at the other.

Their first category, instructional, deals with drill and practice software and is concerned with a largely teaching, or didactic approach. The key feature of this type of program is that it takes the form of a series of structured questions and answers, with the learner being very strongly led by the computer. The user answers a closed question and according to the accuracy of the answer that is given the computer then provides a further question. If the answer is correct then the software will provide positive feedback and will move the user to the next level; if the answer is incorrect then the program will provide further practice and reinforcement at this level until the child can demonstrate competence. This is very much an objectives-based model of learning, whereby the software is designed to reinforce and extend work related to a narrow objective.

The second category is termed revelatory and is concerned with a learning, rather than a teaching approach, and mainly involves educational or adventure games and simulations. The main difference between software in this, compared with the previous category, is that although the user is still led by the computer, they now have to make key decisions to proceed through the program. In the case of an educational or adventure game, they will have several options, and the software will be sufficiently sophisticated to allow the user to follow a particular strand through to a logical conclusion. The main focus with this category is that the software will provide a significant degree of structure, where real situations will be modelled through the varying of external conditions. As far as primary science simulations are concerned, this might involve simulating a science experiment that would be too dangerous or too expensive to carry out for real, or be too slow, such as illustrating how a plant or flower grows by providing a speeded-up video clip over the Internet or from a CD-ROM.

The third category is conjectural, sometimes referred to as emancipatory, where the computer is being used as a tool to learning by providing a framework within which the children can work. This is probably the category with which we are the most concerned here, in that it allows the power of the computer to be harnessed in order to assist and extend the way that children learn primary science. This is where application software is used such as word processors, databases, spreadsheets and desktop publishing packages to enable the child to sort, search, retrieve and graph information, or write up and present their scientific findings. So, in this particular context the child uses the computer to discover patterns in sets of data, to draw conclusions and engage in the higher-order thinking skills that ensure that the computer is being used in an appropriate, value-added way.

The fourth category, that of open-ended use, is possibly the most intriguing, and given the advances that have been made with educational ICT in the last twenty years or so, possibly the most challenging. This puts the user firmly in control, where the computer is being entirely led by the learner, where to all intents and purposes the child is engaging in programming. To a certain extent, there can never be a truly open-ended use, or even open-ended software, as all hardware or software has some limitations and constraints built into it. However, the concept that children can be in complete control of the computer certainly exists. We have already mentioned that we would not normally advocate that children, or indeed their teachers, directly program a computer, but in order to achieve other relevant outcomes, this can be done as a secondary activity.

As far as primary science is concerned, this would most obviously include programming through control technology where the child inputs commands into the computer and the outputs perform a series of operations on components such as lights, buzzers or motors. Typical examples of this might involve developing a traffic light sequence or making a buggy move backwards or forwards, possibly even at different speeds. This is discussed in greater depth in Chapter 6, 'Control technology'. However, a more recent and potentially exciting development is that of web page authoring. Although there are many web sites that can be visited to support the teaching and learning of primary science, little thought has so far been given to children designing their own web sites as a means of presenting their own scientific ideas or discoveries. In order to produce a web page whilst using a specific web authoring package such as *Microsoft Front Page*, *Netscape Composer*, *Claris Home Page* or *Macromedia Dreamweaver*, the user will be dragging, dropping, selecting and composing in exactly the same way that they would if they were using one of the more familiar word processing packages or desktop publishing packages. However, they are effectively writing in HTML (Hypertext Mark-Up Language), which is the main language of the Internet. Indeed, the user can easily click on an option from a drop-down menu to view what they have created written in HTML and although it will not mean a great deal to a child this is not the point. The point is that the children have designed a web page in order to communicate their scientific ideas and understanding. Once this has been posted onto the World Wide Web other Internet users can access their pages and can then feedback their own thoughts and ideas via email.

The kind of experiment that lends itself to this style of ICT, could involve even young children in some original research. For example, we often hear it said that bath water will drain out of the plug-hole in one direction – clockwise in the northern hemisphere, and therefore anti-clockwise in the southern. However, do we know this for a fact or is it conjecture? Using the ICT described above, children from all over the world could settle this argument once and for all!

Although Kemmis, Atkin and Wright's classification may by modern standards appear slightly dated, we believe that it provides a clear framework in which to think about the type, place and purpose of a range of educational software when supporting primary science. Modern advances in software may have ensured that some programs do not fit easily or comfortably into one or other of the categories, but this will also depend upon how that piece of software is being used. If a teacher is using a spreadsheet such as *Number Box* to model a particular condition, and get the children to pose 'What if ...?' types of questions, then this falls into the 'revelatory' category as it is a 'learning' type activity, where the child is expected to answer questions and make decisions based upon these questions. If the teacher is using *Number Box* to record the growth of plants then this will clearly be using the computer as a 'tool' to learning, so will be emancipatory. More recent types of software or ICT such as the Internet and email, which did not exist in the 1970s, would fit into the 'emancipatory' category, as the computer is allowing the user to have the freedom to communicate with others in a way that was previously impossible.

Conclusions

We have seen in this chapter that when using ICT to support the science lesson, the teacher has some key decisions to make. Assuming that using a computer is the most appropriate means of achieving the intended learning objectives in the first place, the teacher has to decide how the computer will be best employed to achieve these. Different teachers will have different teaching styles, and different groups of pupils will have different learning styles. Depending on which role the teacher wishes the computer to perform, these are factors that need to be taken into consideration when selecting software to support the science lesson. If a teacher has a largely didactic style of teaching, then the children will be used to being taught in this way, and as a consequence will be used to learning most effectively in this way. If the teacher wishes to engage the children in some kind of reinforcement activity, then the ICT employed may involve some kind of practice activity, which in turn may involve drill and practice software such as a game or simulation. For all of the limitations of this type of software, the user gets plenty of practice in dealing with a particular concept or idea. The computer will never get bored or frustrated, and will be able to replicate similar conditions every time that it is used. Assuming that the software is sufficiently sophisticated, it enables the teacher to set the conditions via an options menu, ensuring that the level is appropriate to the abilities of the user and relevant to the intended content.

Above all, no matter how sophisticated the software and how good it is, under no circumstances should it ever replace practical science work, which as we have said before, is a fundamentally important aspect of good primary science teaching. Indeed the most powerful use of ICT occurs when it is supporting these practical activities through the use of application software such as databases and spreadsheets. These enable the computer to be used as a tool to extend children's scientific learning through the ability to handle a wide range of different types of data quickly and interactively. Taking this idea one stage further, when the child is put in complete control of the computer, such as when using control technology, the ICT becomes the science experiment itself. So when the child is using ICT as a tool to sort information gathered away from a computer, the software is employed to manipulate data from elsewhere. But when the computer is being programmed to make a set of traffic lights run through a sequence, or when a buggy is motorised and made to move backwards and forwards across a floor, this open-ended use becomes the practical science. Other forms of ICT can then be used to present the findings. A digital camera may record the children programming, and this can then be inserted into a word processing document as part of a *PowerPoint* presentation, or the actual program listings that the children have written can also be exported into these applications.

However the teacher chooses to use ICT as part of the science lesson, it is important to remember that the software that is chosen is a crucial factor in determining the outcome. The overriding consideration remains teaching quality – the software may be the best available, but if it is not used well, or even correctly, then an important opportunity will be lost. It is quite easy to select a particular program simply because it purports to offer the children work in a specific subject or area that the teacher wants covered. It is important that teachers take the same critical stance when choosing software that they would take with non-ICT resources, for both Science and other curriculum areas. As we have discussed at length throughout the chapter, the program must ensure that it does exactly what the teacher wants it to do, and that it will ensure that the intended learning outcomes are met. This is fundamental. It must teach exactly what the teacher wants taught. Nothing can replace good teaching, but this can be affected by the use of poor or inappropriate software. Using the most appropriate software in the correct way can provide unparalleled learning experiences for the children.

The following chapters detail the different types of applications that can be used to enhance the teaching and learning of primary science, including specific programs and case studies to illustrate the kinds of activities that can be completed with children in both key stages of the primary phase.

References

Kemmis, S., Atkin, R. and Wright, E., *How do Students Learn?* Working papers on Computer Assisted Learning, Occasional Paper No. 5, Centre for Applied Research in Education, University of East Anglia, 1977.

Squires, D. and MacDougall, A., *Choosing and Using Educational Software: A Teachers' Guide*, Falmer Press, London, 1994.

British Educational Communications and Technology Agency (BECTa), http://www.becta.org.uk/technology/software/curriculum/evaluation1.html, accessed June 2002.

Chapter 3

Databases

The use of ICT allows children to display their science in a clear and concise form. Nevertheless it is the practical work that the children do that is of the paramount importance, and must be the essential precursor to any ICT. Possibly no other computer activity displays this more than the use of databases and spreadsheets. This chapter will concentrate on databases and Chapter 4 will examine spreadsheets.

As has already been discussed, one of the principal features of a computer is the ability that it possesses to handle vast amounts of different types of data quickly and in a number of different forms. It is this capacity and range that is best exploited by a database, a program that allows vast amounts of information in alphabetical, numerical or alphanumerical form to be stored, retrieved, sorted, searched and interrogated in a structured and organised manner. For example, three megabytes of disk space contains the equivalent of 12,000 index cards worth of information, and a 40 megabyte hard disk can store 160,000 records. However, unlike the card system of a previous generation, any piece of information that is stored in a database can be retrieved in a range of forms immediately and can then be reorganised into any number of different ways, including being displayed as graphs or charts. A primary school database package enables the science teacher to harness this power and flexibility.

As we have already stated, in the early 1980s when computers were first introduced into some primary schools, the software was very limited both in quantity and quality. In the personal experiences of the authors, all that could be done at that time was to list collected data and express it in very simple pie charts and block graphs. These were almost inevitably in black and white or shades of grey. Although this was about all the computer could manage at this stage, it did at least emphasise the importance of the original collection of data, and that it was the time spent by the children over their project that was all-important. This has not changed. The range and type of software may have improved out of all recognition, but nevertheless it still remains the case that it is only when children themselves use the software to collate, sort, search, interrogate and analyse their own results, that the process is of any real value.

Different types of databases

Although some authorities list as many as five types of database, it is our contention that only three are suitable for use in the primary school. These are:

- Free text database, as used to search for information on the World Wide Web or a CD-ROM. This is where the children use the 'search' function of the web page or software to find specific information. In the case of the former, this might involve using a search engine such as *Ask Jeeves*, *Yahoo!* or *Google*, whereas in the case of the latter it will be the search function that is an integral part of the CD program. Whichever is employed, the search strategies are similar. So, for example, if a user wants to find out about electrical circuits, typing electrical+circuits will find all online references to electrical circuits. The insertion of the '+' sign is crucial as it is this that ensures that all references found do not include every single web site that include the words 'electrical' and 'circuit' in any context. Including a '-' sign here will produce all references to 'electrical' but none on electrical circuits.
- Branching or binary tree database, a hierarchical branching database which allows information to be retrieved through the use of questions that require a simple yes or no answer. This database asks the children to describe an object so that it can later be identified by answering a sequence of simple questions. The database may already be set up for the student to follow. For instance, it may be specially written to identify living things such as insects or common trees. Of more interest are the 'blank' databases that allow the pupil to choose what they want to use it for. The structure of the database is there for them, the pupils need only to fill in the necessary questions that will eventually lead to the identification of their chosen objects.
- Random access database, which not only stores the data but enables refined systematic searching and interrogation. This is the type of database that will be familiar to many teachers. Although they may have only used it for the most simple of topics, such as recording the hair and eye colours of their class, they will have found that even with simple projects like this how versatile this kind of database can be. The children would have been able to ask the computer to list the children with, say, fair hair, and match these to the children with blue eyes. This kind of project, which can be used with the youngest of children, requires only a fraction of the potential use of this type of database.

There are many different database packages that are specially produced and are available to primary schools. All conventional databases use the same basic structure. For instance, in a random access database a whole topic such as 'birds', 'minibeasts' or 'ourselves' forms a file, and is saved in the same way as any other program application file. Each individual subject within the file is referred to as a record, similar to one traditional record card that contains the specific information of one kind of bird, insect or person. Each item of information on the record is contained in a field, which is a category of information such as habitat. Although many modern programs allow the user to create their own databases, the teacher needs to map out the structure in advance as this can be difficult to change once created. Teachers may need to devise a particular structure for their database depending on the type of information that is held, and ultimately, what is to be done with it. Indeed, selecting an appropriate structure and database is an important skill in itself when using databases for science at Key Stage 2. However, it has not always been appropriate for primary age children to construct their own databases from scratch as there are simply too many variables to take into consideration. Certainly, they could identify file names and appropriate records and field names, but actually devising the structure was too complex. However, recent programs that have been devised specifically for use in the primary school, such as *First Workshop* and *Information Workshop* which form part of the *Black Cat Toolbox*, published by Granada Learning, contain powerful functions that enable files to be constructed simply and easily with a minimum of adult help. The teacher can then ask the class to identify, collect and enter the data and subsequently interrogate this information. The focus of the overall activity is to get the children to use the higher-order scientific skills of data collection, preparing the data for entering into the database, entering the information into the database and the subsequent activities of sorting, searching, retrieving and interrogating the resulting information.

Using the power of a database

Powerful search and sort functions are an important part of the use of databases in primary science, for they represent the principal key to accessing the higher levels of understanding that gives the use of a database its value-added component. The ability to look for and find quickly different types of data, often buried quite deeply within a huge amount of information, is valuable enough. To then display it on the screen, look for relationships, similarities, differences and patterns, either by trawling through the data or through more visual means by the production of a graph or a table, is extremely powerful and one that accesses new levels of learning. For example as part of a topic on 'ourselves', a

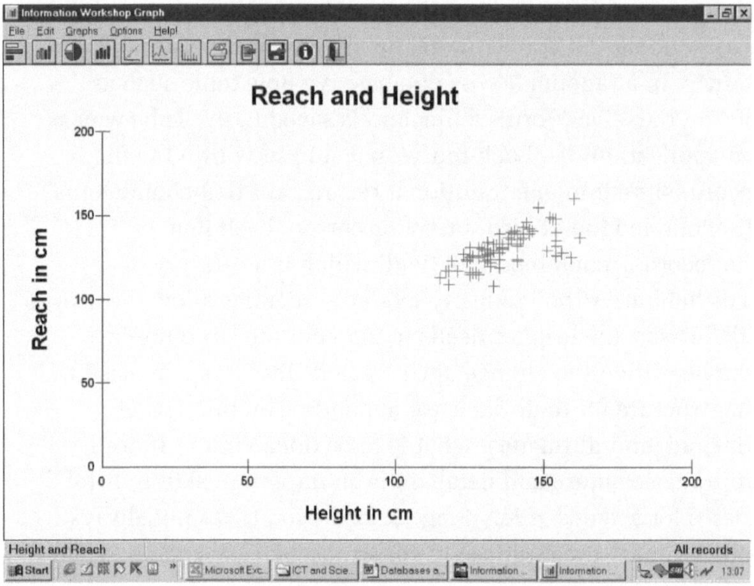

Figure 3.1 Scattergraph produced by *Information Workshop*.

database can be interrogated for the number children who have brown eyes, fair hair, favourite foods or the number and types of pets. The computer can search the file, and display any records of children who meet one, or even all of these criteria. The children might look for relationships between two sets of data, for example height and armspan. This information can be ordered and then displayed as a scattergraph on the screen, so that any relationship can be displayed in a very clear and visual way. For example, if all of the crosses are found to be grouped together, it can be said that there is a positive correlation between the two. This is illustrated in Figure 3.1. The grouping of the crosses on the scattergraph means that there is a strong correllation between height and armspan. This is because height is normally only slightly greater than reach.

In this way information that is refined can also be displayed in graphical form. Children produce simple graphs that are not only derived from the basic data, but also from information that has been produced as the result of a quite specific search or sort. Even quite young children can engage in these kinds of searches. According to Ager (1998) children in the Early Years should be able to add data to a pre-prepared database such as those found in *First Workshop* and, using the appropriate level, display this information in meaningful ways. The use of databases and spreadsheets can support the requirements listed in the Early Learning Goals (1999), especially Numeracy and Knowledge, and Understanding of the World. However, when the children reach Key Stage 2 they should then be able to interrogate the database in greater depth to identify new patterns of information. It is at this point that the children should become aware of the main advantages of computerised databases. It is also at this stage of their development that they will need to be aware of how information can be obtained from the World Wide Web and CD-ROMs, and of how this type of database search relates to the processes involved in obtaining that information.

Teachers will become aware that whilst building a database a certain set of skills will be required but interpreting it will require a different group of skills. The children will need to identify and search different sources of information and make value judgements as to which are useful, relevant, valid, accurate and which are not. They may have to plan and put together a search strategy, which in turn will involve the framing of useful questions. When looking for information they will need to use keywords and operations such as AND, OR and NOT. For example, this might be looking for a simple, single relationship, or multiple relationships. On the 'ourselves' file, this might involve looking for those children who have blue eyes AND fair hair. This interactivity, one of the functions of ICT, allows rapid and dynamic feedback and response, and thus does not detract from the development of the higher-order skills. At whatever level, for any science project, using a database will require the children to collect their raw data, prepare it for entry into the database and organise it in such a way that it can be stored and retrieved later for interrogation, interpretation and if necessary, correction or sorting into a different order such as alphabetical, chronological or numerical. When retrieved, the children will have to learn how to interpret the information in a meaningful way. As a direct consequence of this the children will learn to consider the validity, reliability and reasonableness of outcomes. One advantage of using a computerised database is that any file can be saved and later used in a number of different ways. This can be carried out with great speed. Records can be added or deleted quickly, as can the information contained within them. This information only needs to be stored once, and can then be organised in a number of different ways. Records can then be linked together as required and if necessary the entire structure can be changed. If the files were paper based they would have to be updated simultaneously if the same information was stored in several different places, a difficult and time-consuming process.

As we have suggested earlier, it is the branching database and the random access types that are the most valuable in a primary science context. When used properly, with both these databases, the child is required to devise and pose key questions, based on the scientific work that they have undertaken. They should be encouraged to think what kind of information they would want from a database. In a random access database, they might, for instance need to know how many different types of animal live in a certain habitat, or how many days over a given period had no rain. By considering these outcomes, they will be made aware of the value of the database, and at the same time realise the relevance of the raw data that they have collected.

Branching databases

A study of materials can provide the opportunity to construct a branching database. The teacher will need to ask the children to think of suitable questions with yes or no answers that they can ask about the materials they are studying. They might ask questions such as, 'Does it bend?' or 'Is it transparent?' After each answer, the children need to decide whether they have sufficient information to identify the material, or if they need to ask another question. This continues until the 'choice' of material is narrowed down to a single example. In order to effectively construct a branching database, a different way of thinking is required. Previously in this book we have discussed the importance of using open-ended questioning as part of the children's science investigations that seldom require a straightforward, closed yes or no answer. However, in this case this is exactly the approach that is required, as the children need to approach the problem from a different perspective, or as McKenzie (1997) states, "The question is the answer", as the exact phrasing of the question becomes all-important. Therefore, in order for the children to fully appreciate this and to get into the correct mindset it will be necessary for them to map out their decision trees. This can be done by either drawing out a draft plan of the tree on a large sheet of paper, or as Wake and Atkins (2000) suggest, by laying physical objects or pictures and photographs on the floor, with cards and arrows to model the structure. Either way, it is necessary, particularly with younger children, to have a practical and 'real' stage first. It will also be helpful for them to look at and interrogate a completed decision tree, perhaps a commercially constructed one that has come as a sample file with the branching database software. This will enable the children to make the conceptual link between the abstract computerised database and the real data that they are going to use.

To carry out this kind of materials project it is better for the children if they have ready access to a computer located in the classroom. They will have been learning about the properties of various substances, and have a collection of them in the room. As they build up the database the children will be able to turn to the collection to remind them about the questions they should ask. They will have tested the materials for such properties as hardness, permeability, flexibility, and malleability. They would have looked to see if they were shiny or dull, transparent or opaque. They might have even tested them to see if they would float, or if in their opinion the material was 'heavy' or 'light'. They should have also tested them for electrical conductivity.

Ask the children to record all this information in their notebooks. The building of this database can be carried out as a group activity. If one group builds this database, other groups should have the opportunity to build other branching databases at a later stage. However, it is possible for all the children to have an input into the making of the database, with several (or even all) of the children taking it in turns to work at the computer.

All the children can take part in deciding what kinds of questions to ask the computer. As we have explained these must have the answer 'yes' or 'no'. This is not as easy as it seems, particularly for children, especially as many of them may think of a negative answer as wrong. It is not just, as is often the case of a branching database about animals, a question of looking at similarities and differences. These are of course very important, but children will find it more difficult to pose the right question about the properties of a material as they would about the number of legs on an insect. These simple questions, with their correspondingly straightforward answers, are ironically not always the type of question that many teachers would wish to ask. More often, in a science lesson or project, the imaginative teacher will be looking to devise and ask searching and open-ended questions of the 'why', 'where' or 'how' variety. However, as we have mentioned elsewhere, the first requirement of observation in science is for the children to discover what is actually present and what is happening to it! The questions that they are required to answer for the branching database are of this kind. 'Are their six legs?' or 'Are there three parts to the body,' require simple yes or no answers, obtained from the straightforward observation of any minibeast. Questions that might be asked during a study of evaporation, such as 'Where do you think all the liquid has gone, and in what form could it be?' are not for a branching type database.

We have pointed out in the chapter on planning that it is necessary to ensure that the standard of science matches the level of the ICT. Although it is simple to use, a branching database is a sophisticated tool for learning. It should be used as such. Children should be encouraged to build up a bank of information so that the database can become a permanent reference source for the rest of the school. In this case, as new materials are studied, perhaps by other year groups, so the children can refer to the database to see if they can identify them. If they cannot, then they in turn can add information to the branching database.

Random access databases

With a random access database, information is again fed into the program, but on this occasion, not as a key to identification, but with a view to collecting data, so that it can later be sorted, collated and recalled when appropriate. It is this use of the database that has a truly educational value. Building the database is only one factor in the process. Only when children realise how the information can help their learning, decide upon what intelligent questions to put to the computer to obtain that information, and then choose in what way it should be displayed, can the value of the work be finally realised.

As we mentioned earlier, even this kind of database can be built and used by all ages of children. As long as the correct level is found, this is a valuable activity for even the youngest of Key Stage 1 children. Two databases that allow for a choice of levels of difficulty are *First Workshop* and *Information Workshop* which are both part of the *Black Cat Toolbox*. *First Workshop* is designed for use by children in the Early Years and enables them to design and complete their own databases. This process is illustrated in Figures 3.2 to 3.8.

- Step 1: The children open *First Workshop* and click on 'Start a new topic'.
- Step 2: The children select 'Choose a topic' and click on 'Next'.
- Step 3: The children select which fields they want in their database and then click on 'Finish'.
- Step 4: The children are presented with a blank record sheet where they enter their details next to the appropriate field name.
- Step 5: When certain fields are selected which require correct spellings but only have a limited number of options, a menu appears. The child selects the option they require and this is pasted next to the field name. These are known as keywords.
- Step 6: This is repeated with the 'favourite sport' option.
- Step 7: The completed record for James.

Information Workshop is more suitable for children at Key Stage 2. Similar in layout to *First Workshop*, it allows for differentiation by providing three different levels of sophistication.

We have discussed the structure of databases in an earlier paragraph and how, when creating the database, children will need to be conversant with the titles given to these various entries, namely *file name*, *field*, and *record*. One of the most popular uses for a database is for the study of animals, either for general reference or for a more specific project carried out in the field. For the latter, this may be a survey of the school grounds. The children may observe a considerable number of small animals, and depending on the level of work that they do, will

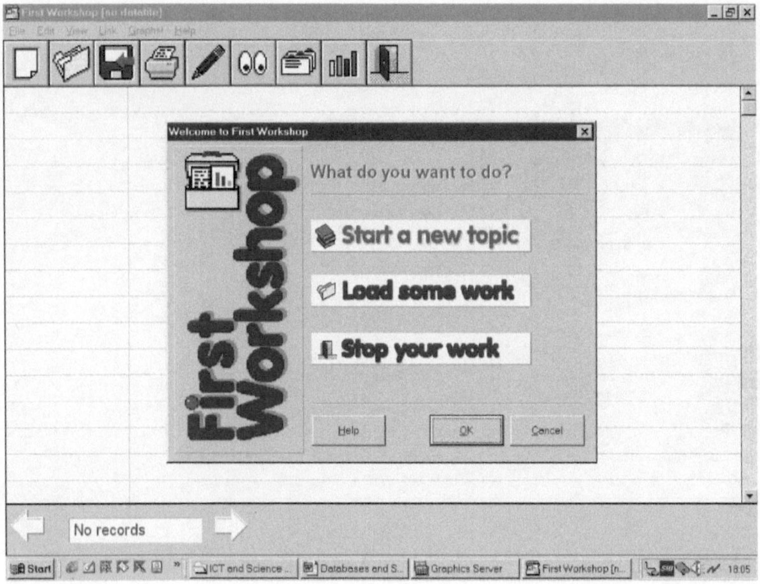

Figure 3.2 Step 1.

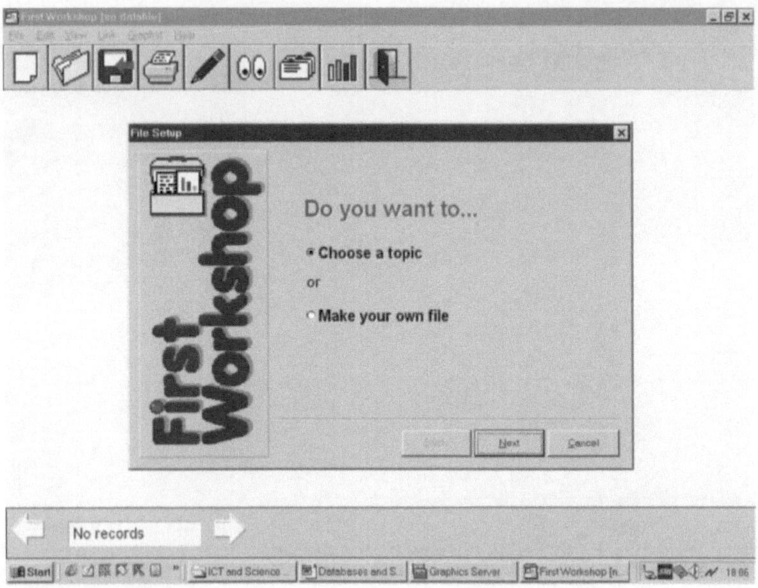

Figure 3.3 Step 2.

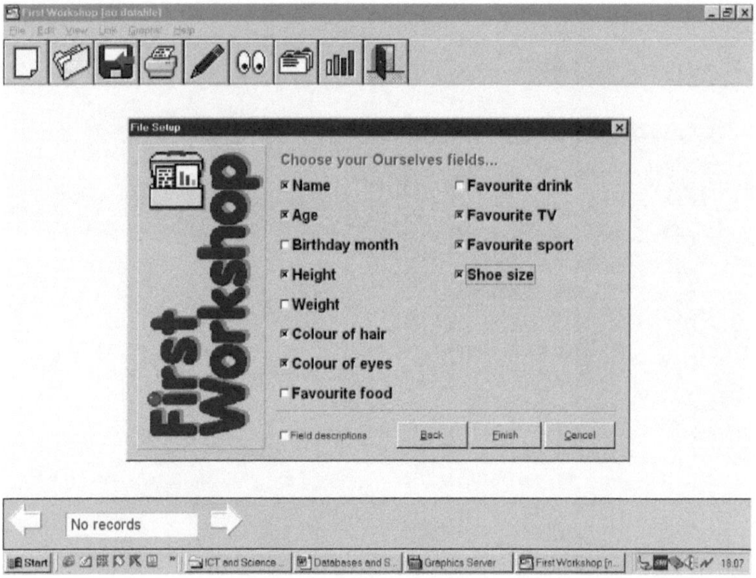

Figure 3.4 Step 3.

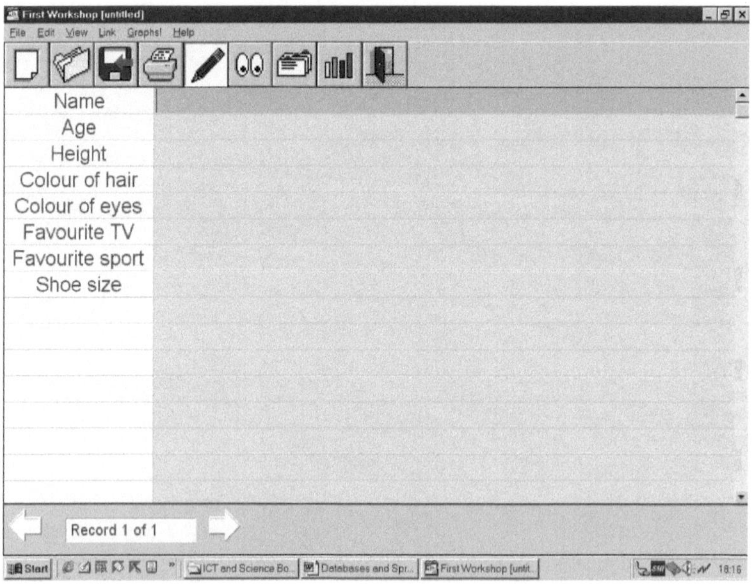

Figure 3.5 Step 4.

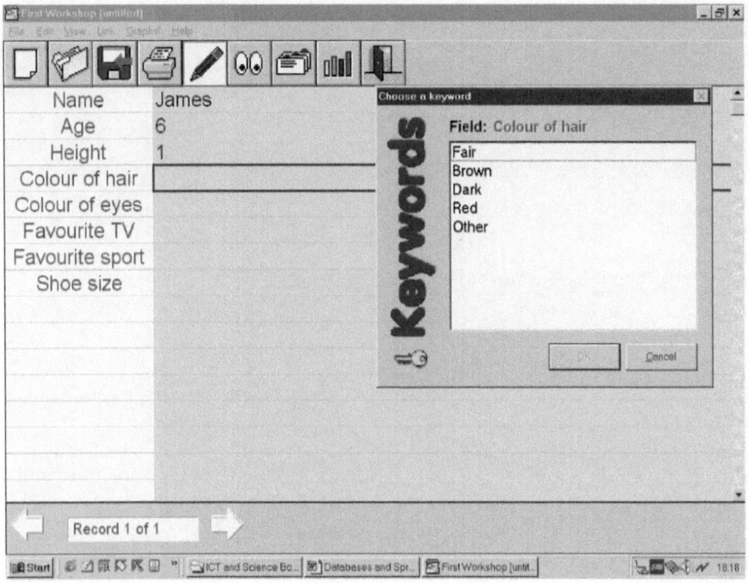

Figure 3.6 Step 5.

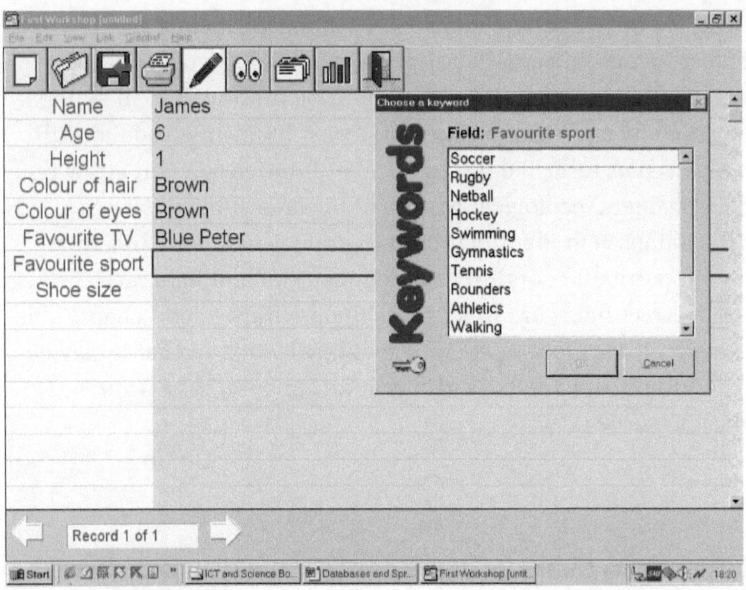

Figure 3.7 Step 6.

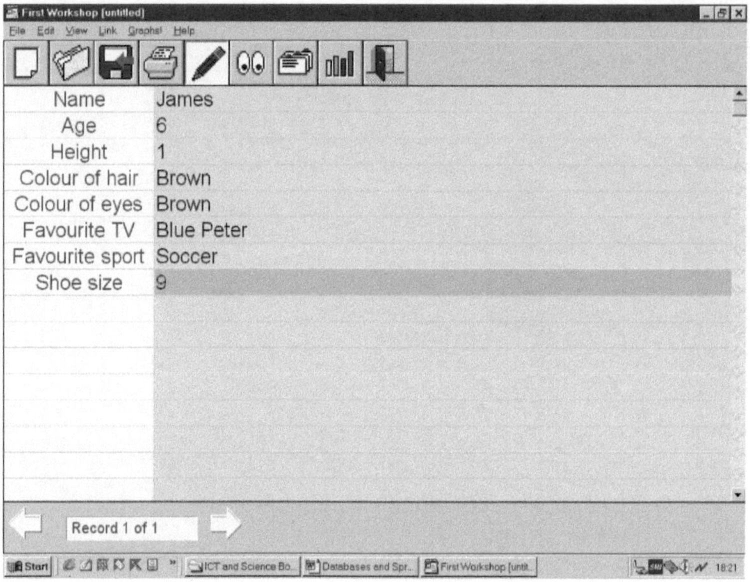

Figure 3.8 Step 7.

have already identified them, perhaps by using a branching database. What purpose, then, has this new database? Whilst it can be used for identification purposes, it cannot be overstated that the main purpose of this kind of database is to help children come to a variety of conclusions about what they are studying, and not just identifying and listing what is present. Hence, once the file name has been entered, the various fields should include ecological factors, such as where the animal was found (e.g. under stones, in rotting wood, or full sunlight as against partial or full shade). This kind of information will allow the children to build up a general ecological picture of the area studied. Hence, children can ask the database to identify groups of animals that may live in specific habitats, with particular conditions of temperature and light, and perhaps even, if the correct fields are used, place them within a food chain. This kind of data can also produce more interesting graphical work, in that it may be necessary for the children to decide not only what information they want shown, but how best to show it.

Case studies

For this study we must thank Class 7 of Bluehouse Community Junior School, Basildon, Essex, and their teacher, Karen Smith, for all the extra hard work that she carried out on our behalf.

Class 7 consisted of thirty Year 5 children. They were making a study of animal types, their characteristics and habitats in keeping with the National Curriculum. They had begun the project by considering those animals, i.e. pets, with which the children had first-hand knowledge. A recent visit to the school by a herpetologist had given them an experience of more exotic types. From this introduction they were able to use a variety of visual aids to increase their knowledge of the more common of the world's fauna.

They produced a variety of diagrams, writings and short booklets about many of the different animals studied, as well as a vivid wall display in their classroom. It was then decided to use this information to produce a database for future reference.

Together with their teacher, they first designed an identification card for the animal. It was decided that instead of having five specific animal characteristics per card, two would be biological, whilst the other three would be environmental (i.e. ecological). Hence, what was to become the 'field names' for the database were:

- Habitat
- Area of the world
- Young
- Feeding
- In danger of extinction?

Each characteristic was given a variety of sub-headings, so that, for example 'Habitat' had a choice of sea, river, pond, land, farm and home. All these would become the 'Keywords'.

Each child was asked to pick an animal, and fill in the details on a card by colouring in the relevant spaces. A blank example of one of these is illustrated in Figure 3.9.

Although these cards were later to be used to feed data into the computer, the teacher first used them to show the children how a database works. She first asked the class to organise themselves in alphabetical order according to the name of the animal on each of their cards. The children were very good at doing this, and although several animals had names beginning with the same letter, had little difficulty in finding their correct place in line. When this was completed, the teacher then used the cards to show how a database works. She read out various words from a chosen card such as 'Habitat – River, Area of world – Asia' and asked those children whose cards did not fit the description to

Name

Name of animal ....

Habitat

sea	river	pond	land	farm	home

Area of world

oceans	Europe	Asia	Australasia	Africa	North America	South America	Arctic	Antarctic	Lots of places

Young

Has live babies	Lays eggs

Feeding

Herbivore	Carnivore	Omnivore

In danger of extinction?

Yes	No

Figure 3.9 Card used by the children to record their data.

sit down. When only one child remained standing, the animal named on that child's card was identical to that on the card chosen by the teacher. Although the children found these games interesting and took part in them enthusiastically, they nevertheless took some time to complete. It was then pointed out that a computer would do this almost instantaneously, and with many more examples than the thirty represented by Class 7.

The next lesson took place in the computer room. This is also the school library. There were enough computers for most of the class to share one between two, with a small group working on further research in the library, whilst they waited their turn at the computers. The class teacher was the only adult present, but the children very quickly settled down to create their files. Using *Information Workshop* which was already installed on the computers, the children chose the Red (most advanced) level. They named their new file 'Animals', with their class number as the author. They then completed each field, and filled in the appropriate keywords. It took about two sessions for all the children to complete the database, each one entering the details of over thirty animals. This is illustrated in Figure 3.10.

For organisational reasons these two sessions had to be carried out on a class basis. However, after these formal sessions, the children were able to work in small groups and so explore the nature of the database, and its potential for

No.	Name of animal	Habitat	Area of world	Young	Feeding
1	Tiger	land	Asia	Has live babies	Carnivore
2	Sheep	farm	Lots of places	Has live babies	Herbivore
3	Shark	sea	Lots of places	Lays eggs	Carnivore
4	Rabbit	land	Lots of places	Has live babies	Herbivore
5	Porcupine	land	Lots of places	Has live babies	Herbivore
6	Polar Bear	ice	Arctic	Has live babies	Carnivore
7	Penguin	ice	Antarctic	Has live babies	Carnivore
8	Otter	land	Europe	Has live babies	Carnivore
9	Orangutan	land	Asia	Has live babies	Herbivore
10	Martial Eagle	land	Lots of places	Lays eggs	Carnivore
11	Lynz	land	Lots of places	Has live babies	Carnivore
12	Lion	land	Africa	Has live babies	Carnivore
13	Indian Elephant	land	Asia	Has live babies	Herbivore
14	Horse	land	Lots of places	Has live babies	Herbivore
15	Hedgehog	land	Europe	Has live babies	Omnivore
16	Grizzly Bear	land	North America	Has live babies	Herbivore
17	Grey Heron	river	Lots of places	Lays eggs	Carnivore
18	Great White Shark	sea	Lots of places	Has live babies	Carnivore
19	Giant Panda	land	Asia	Has live babies	Herbivore
20	Eagle Owl	land	Lots of places	Lays eggs	Carnivore
21	Dolphin	sea	Lots of places	Has live babies	Carnivore
22	Crocodile	river	Africa	Lays eggs	Carnivore
23	Chimpanzee	land	Africa	Has live babies	Omnivore
24	Budgerigar	home	Australasia	Lays eggs	Herbivore
25	Brown Bear	land	Lots of places	Has live babies	Carnivore
26	Blue Whale	sea	Lots of places	Has live babies	Carnivore
27	Beaver	river	Lots of places	Has live babies	Herbivore
28	Arctic Fox	ice	Lots of places	Has live babies	Carnivore
29	Green Turtle	sea	Oceans	Lays eggs	Herbivore

Figure 3.10 Table produced in *Information Workshop* containing all of the information entered by the children.

extracting information. They experimented with the various icons along the top of the screen, and soon found that they were able to obtain records for the various characteristics entered into the record. Hence, for example, they could soon obtain data that showed how many of their animals in Africa live in rivers (see Figure 3.11).

As well as becoming skilled at using the database, the children also found that they were able to list their information under various headings, and represent it in various graphical forms. All the children, at whatever ability level, worked with enthusiasm and interest. All could manage the basic skill levels necessary to complete the database, and many of them showed obvious potential to reach much higher levels in the future.

Figure 3.12 shows a graph produced in *Information Workshop* to illustrate the distribution of animals in the survey around the world. Figure 3.13 shows the same information displayed as a 3-D pie chart; and Figure 3.14 shows the information sorted and displayed by diet.

By using the information gathered from their science for their computer education, the children not only reinforced this knowledge, but found an added interest when presented with the ICT tasks. Before the advent of the computer (a time clearly remembered by both of the authors), the children would have been

No.	Name of animal	Habitat
3	Shark	sea
18	Great White Shark	sea
21	Dolphin	sea
26	Blue Whale	sea
29	Green Turtle	sea
17	Grey Heron	river
22	Crocodile	river
27	Beaver	river
1	Tiger	land
4	Rabbit	land
5	Porcupine	land
8	Otter	land
9	Orangutan	land
10	Martial Eagle	land
11	Lynz	land
12	Lion	land
13	Indian Elephant	land
14	Horse	land
15	Hedgehog	land
16	Grizzly Bear	land
19	Giant Panda	land
20	Eagle Owl	land
23	Chimpanzee	land
25	Brown Bear	land
6	Polar Bear	ice
7	Penguin	ice
28	Arctic Fox	ice
24	Budgerigar	home
2	Sheep	farm

Figure 3.11 Table showing animals sorted by their habitat.

expected to record the results of their science projects with pen and paper. They would have completed it well, but it would have taken a lot of time. Class 7 showed how, once the basic commands of the computer had been learnt, these records could be made quickly available. It is in this way that ICT becomes an integral part of any good science project.

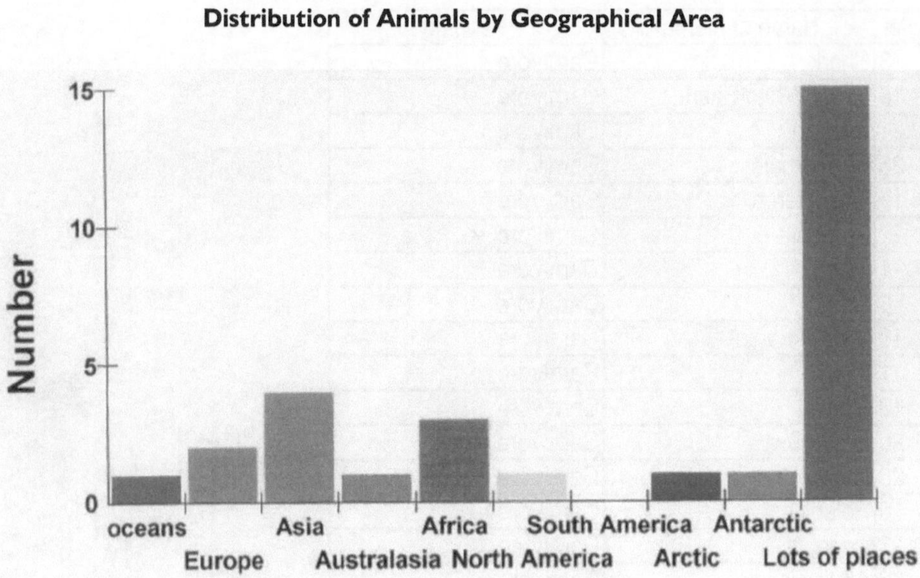

Figure 3.12 Graph illustrating the distribution of animals in the survey around the world.

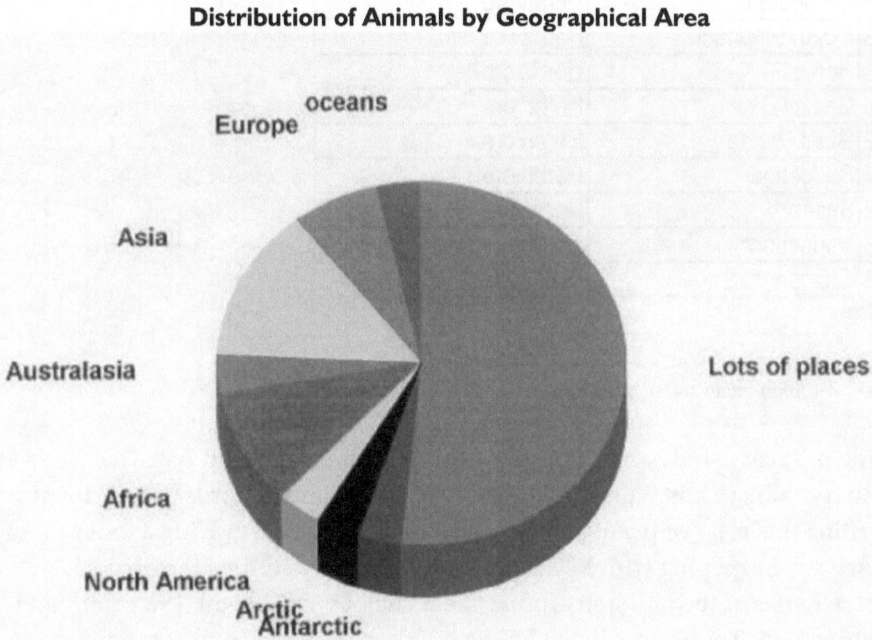

Figure 3.13 3-D pie chart illustrating the same information as Figure 3.12.

No.	Name of animal	Feeding
3	Shark	Carnivore
18	Great White Shark	Carnivore
21	Dolphin	Carnivore
26	Blue Whale	Carnivore
17	Grey Heron	Carnivore
22	Crocodile	Carnivore
1	Tiger	Carnivore
8	Otter	Carnivore
10	Martial Eagle	Carnivore
11	Lynz	Carnivore
12	Lion	Carnivore
20	Eagle Owl	Carnivore
25	Brown Bear	Carnivore
6	Polar Bear	Carnivore
7	Penguin	Carnivore
28	Arctic Fox	Carnivore
29	Green Turtle	Herbivore
27	Beaver	Herbivore
4	Rabbit	Herbivore
5	Porcupine	Herbivore
9	Orangutan	Herbivore
13	Indian Elephant	Herbivore
14	Horse	Herbivore
16	Grizzly Bear	Herbivore
19	Giant Panda	Herbivore
24	Budgerigar	Herbivore
2	Sheep	Herbivore
15	Hedgehog	Omnivore
23	Chimpanzee	Omnivore

Figure 3.14 *Information sorted and displayed by diet.*

For the next case study we would like to thank Sarah Cass and Ash Tree Class at Westbury Primary and Nursery School, Letchworth in Hertfordshire. At the time of writing this class of twenty-three Year 2 children were studying a topic about 'Ourselves'. In keeping with Key Stage 2 children they studied the obvious characteristics of the children around them such as age, height, eye colour and favourite food. These were listed by the children on a record card produced by their teacher, who is also ICT Co-ordinator for the school.

Figure 3.15 Pictures created using the 'Faces' screen of My World.

The children were introduced to *My World,* a framework package that enables children to drag and drop items around their chosen environment. They first used the 'People' screen of *My World* so that they could become familiar with the techniques required to use the program successfully. To produce a visual representation of their visible characteristics they used the 'Faces' screen to construct pictures of each other. Some examples of these are illustrated in Figure 3.15.

Once the children had become conversant with this program it was decided to introduce them to the techniques of entering data into a database. Using *Information Workshop* at the Red level, their teacher set up the database for them but the children completed the fields. Although it is appropriate at this age for the teacher to set up the database itself, the children very quickly became highly skilled at entering the data collected on their record card. They were then able to analyse this information to look for relationships between different elements of the data that they had collected. The children completed a record card with a pencil first (Figure 3.16). Then they completed a record card using *Information Workshop*(Figure 3.17)

Age	7
Height	122 cm
Year group	Year 2
Eye colour	blue
Hair colour	blonde
Favourite food	Chips pizza
Number of brothers	5
Number of sisters	3
Type of house	semi-detach-ed
Number of rooms	9

Figure 3.16 Record card completed with a pencil.

Name	Daisy
Age	7 years
Height	133 cm
Year group	2
Eye colour	brown
Hair colour	blonde
Favourite food	jelly
Number of brothers	0
Number of sisters	3
Type of house	terraced
Number of rooms	9

Figure 3.17 Record card produced using *Information Workshop*.

Conclusion

These case studies represent just two examples of how teachers have used ICT to enhance the science teaching in their classes. The children could have carried out their science investigations without going near a computer, and the ICT element could have been done from a set of fictitious information. However, using a genuine science project, studied at first hand, and combining it with the ICT, not only added further interested and motivation for the children, but it brought an authenticity to the work that would not otherwise have been present. The power of the computer was employed in several key ways. In the first case study, the children used the computer to sort their ideas by producing record cards that provided a sharp focus to their investigations. Then they used the database to construct a clear, yet detailed record of animals' feeding habits and their geographical distribution. In the second case study, the children produced images of their own faces using the framework package *My World*, then entered their data into *Information Workshop*. This integrated task also ensured that the children were focused on higher-order skills work such as analysing information rather than becoming bogged down by drawing faces or spending hours hand-drawing a range of graphs, charts and tables. Although these skills are important, it is not necessary to produce these every single time, so the children can concentrate on the 'Why?' and 'What if...?' questions rather than never being allowed to access these higher levels of thinking due to spending so much time on tasks of 'drudgery'.

Bibliography

Ager, R., *Information and Communications Technology in Primary Schools: Children or Computers in Control*, David Fulton, 1998, pp 67–70.

Crompton, R. and Mann, P. (eds), *IT Across the Primary Curriculum*, Cassell, 1997, pp 79–81

Drage, C., *Science: Primary ICT Handbook*, Nelson Thornes, Cheltenham, 2001.

Feasey, R. and Gallear, B., *Primary Science and Information and Communications Technology*, Association for Science Education, Hatfield, 2001.

Harlen, W., *Primary Science: Taking the Plunge*, Heinemann, London, 1988.

Loveless, A., *The Role of IT*, Cassell, 1995, pp 57–62.

McFarlane, A., *Information Technology and Authentic Learning*, Routledge, 1997, pp 153–154.

McKenzie, J., 'The question is the answer', *From Now On* vol 7 no. 2, 1997, online at http://questioning.org/Q6/question.html, quoted in Wake, B. and Atkins, P. below.

Poole, P. (ed), *Talking About Information and Communication Technology in Subject Teaching*, Canterbury Christchurch University College, 1998, pp 70–71.

QCA, *Early Learning Goals*, Qualification and Curriculum Authority, London, 1999.

Wake, B. and Atkins, P., *Asking the Questions: Using Branching Databases with Young Children*, Focus on Science, Micros and Primary Education (MAPE), Newman College, 2000.

Chapter 4

Spreadsheets

What is a spreadsheet?

All good science work involves clear and accurate recording, and this is where the use of a spreadsheet can play an important role. Databases are ideal for manipulating data, especially searching, sorting and displaying the resultant information in a variety of ways. However, there are certain types of data that they cannot easily handle, such as numerical information that may have been collected as a result of a tally chart or similar, such as with a pulse rates topic. This type of information is best displayed in a spreadsheet, because several types of similar information with a quantifiable output can be displayed and graphed or analysed immediately. Like any other computer program, a spreadsheet can be designed for the age and ability of the user, just as much as what it is required to do. Hence, at its most simple level, a spreadsheet can just list a table of results or observations. These could be a list of day-by-day measurements of a growing plant, or series of measurements of the various members of a class of children. However, the spreadsheet has many added advantages. It can model a particular outcome or activity, where children can enter data and find out what happens when some of it is changed (Feasey and Gallear, 2001). Spreadsheets can utilise the basic data to produce further information. For example, they can calculate averages, or in the plant example give actual growth rates. Open-ended questions of the 'what if?' type can be posed to extend children's thinking and consequently their learning, and from this 'imaginary' data, graphs can be produced and comparisons made with either other modelled data, or perhaps even real data. This would be difficult to do with a database, and as a result of this children need to be taught which type of data handling package is the most appropriate for any given task, as the outcome of an entire investigation can be affected by the selection of an inappropriate program. Although the children will need some previous experience of using both types of software before they are able to make a judgement as to which one is the most appropriate, the teacher plays an important part in helping the children to decide which is the most suitable for a given task.

When should a spreadsheet be used?

If the children were recording plant growth, a discussion between the teacher and a Year Two class might look something like this:

Q. 'What type of information have you collected?'
A. 'How much the plant has grown in centimetres'.
Q. 'How are you going to write this down?'
A. 'By measuring the plant every day and writing the height of it next to what day it is.'
Q. 'Which program are you going to use?'
A. '*Information Workshop*' (database)
Q. 'Why? How are you going to use *Information Workshop* to show this?'
A. 'I can make a new card for every day to show how much it has grown.'
Q. 'But how will it show how much it has grown?'
A. 'By drawing a graph.'

Although the child has got a good idea of the outcome, the understanding of the process to get the outcome is incorrect. This would not work, for although the database would show growth it would not show it cumulatively. The whole point of the exercise is to record the total growth, not the growth for each separate day. Even if the child had used one record card but used each field for recording daily growth, there would still be no means of calculating either daily or cumulative growth. However, by using a spreadsheet both aspects could be calculated and displayed easily and clearly. The teacher would need to explain this essential difference to the children – databases search and sort information, whereas spreadsheets manipulate and model numbers. An example of a Key Stage 1 spreadsheet for plant growth is illustrated in Figure 4.3.

In the primary school, children are taught how important it is to make clear and accurate records during any science project. These will often include a description of the experiments as well as why they needed to be done. There may be a series of relevant drawings, and of course a record of the results and observations of the experiments themselves. There may well need to be a list of clear and obvious results that can be tabulated in one form or another. At first sight this may be only what a spreadsheet is – a blank form on which the children can record the results of their science. Whilst it certainly does have this purpose it also enables the user to make much more flexible use of the data once it has been entered. Indeed it has many of the same functions and can perform many of the same operations as a database, but can be too complex to use in this form with primary-aged children. However, if fully understood it is probably the most useful program available.

Spreadsheets are particularly valuable when any mathematical calculations are required. However, young children will need to be introduced to them gradually. The formulae required to make the best use of spreadsheets can be quite sophisticated and are better left until the latter stages of Key Stage 2. However, children should use a spreadsheet to record their results as soon as they can, and make use of its ability to handle data quickly and present it in graphical forms. In using these sheets for recording, children will at least soon learn how to fill in the spaces, or cells as they are called, and be able to alter their size and format. As they become more confident and experienced in their use, they will be able to use the data and the displayed information as a starting point for discussion, and take note of any relationships and trends that appear within the displayed data (Feasey and Gallear, 2001).

As an introduction to the full-blown spreadsheet, teachers should use some of the ready-prepared ones that are available. For instance an important program in the Black Cat Suite, *Number Box,* is a delightful child-friendly spreadsheet program. The teacher can use the simple blank spreadsheet presented to them, or they can click on an option that will give them a range of pre-prepared spreadsheets for several different topics. Whilst some of these are strictly mathematical (i.e. shopping), most are made for science projects. Two of the authors' favourites are 'Temperature' and 'Pulse Rates'. The former has columns for Day (the date), Time, and 1, 2 and 3 (Figure 4.1). These represent three thermometers set in the school grounds. There are sections listing the maximum and minimum temperatures, and cells that will give a reading that shows the average temperature as well as the temperature range. All spreadsheets can give these, but with these 'quick' sheets these will appear automatically.

The 'Pulse Rates' quick sheet (Figure 4.2) has columns for the name of the child being tested, and for the standing, walking and running pulse rates. They will also show the average, maximum and minimum rates for each individual child, as well as a total average, maximum and minimum rate.

Of course these programs should form the basis of a genuine science project. The temperatures should be obtained from three genuine thermometers being used in a genuine ecological project, and the pulse rates should be a part of a

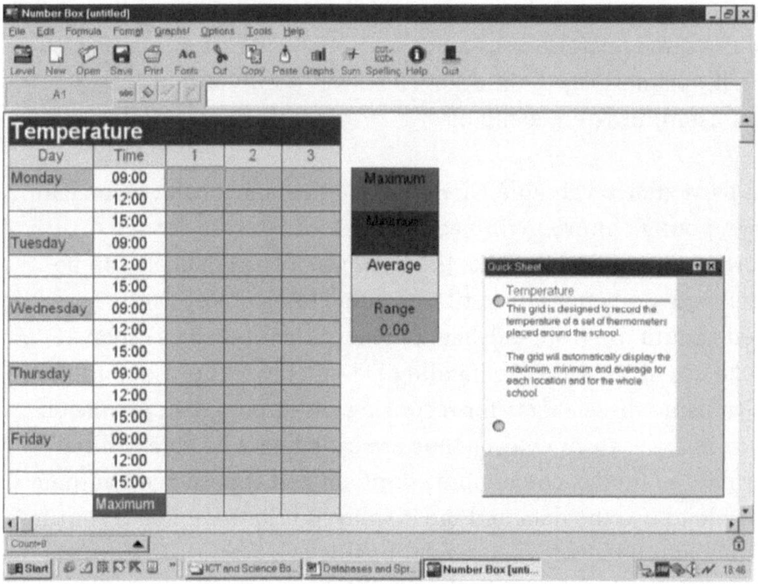

Figure 4.1 *Number Box:* 'Temperature' quick sheet.

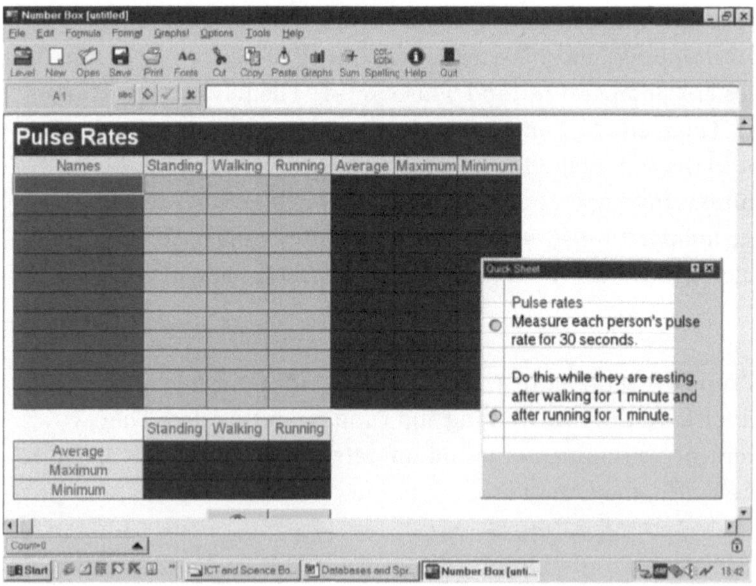

Figure 4.2 'Pulse rates' quick sheet.

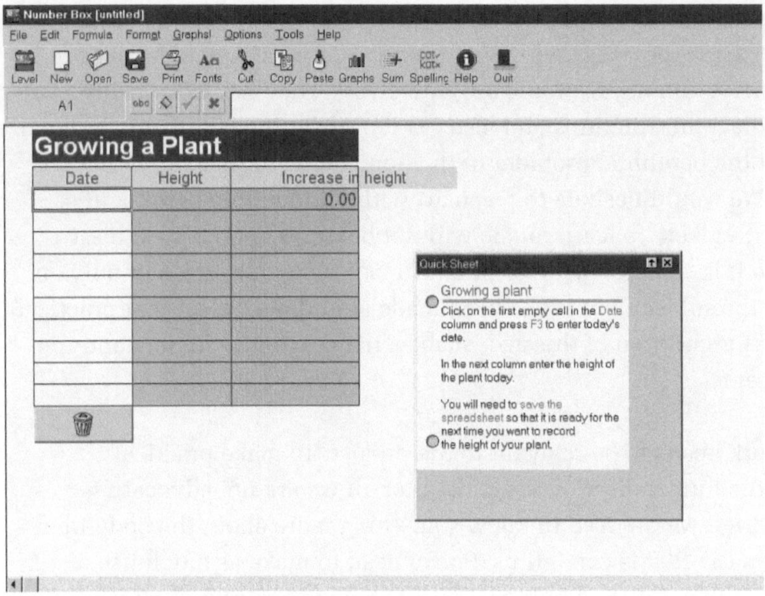

Figure 4.3 'Growing a plant' quick sheet.

practical study involving 'Ourselves'. It is only in this way that children will learn how best to harness these spreadsheets to help them in their scientific studies, and to adapt them where possible to meet their particular requirements. For example, if they use the 'Growing a plant' spreadsheet (Figure 4.3), they will soon realise that plants do not stop growing at weekends! They will need to make some decisions about how to adapt the sheet to account for this. They will also realise, perhaps for the first time that a continuous line graph will give the best representation of a continuously changing record. Hence when presented by the block graph that this program produces, they will soon come to understand that they can superimpose their own line graph onto the print-out, by joining up points at the top of each graph column. This is one of the many examples of how programs like these can add to the decision-making process, which is such an important part of any child's education.

Case study

For this study we would like to thank Class 4/5 at Westbury Primary School, Letchworth, and their teacher Anna Marshall for all their time, effort, and patience, and for allowing us to use their work.

The class has thirty-two children, of a wide range of ability, drawn from both Years 4 and 5. They had been studying the structure and function of the human body in keeping with the National Curriculum requirements – Sc1 Scientific Enquiry and Sc2 Life Processes and Living Things – in what would have been in pre-National Curriculum days, a topic on 'Ourselves'. They had been using suitable books, diagrams, pictures and also CD-ROMs for their research, but had come up against the perennial problem in this topic: how to find suitable practical work. We would hesitate to teach art without touching paint, and children are hardly likely to learn music without having a chance to at least make a noise! So it is with science, for practical experimentation is not only an inherent part of the subject, but it has been made abundantly clear that practical work, especially for children of this age, enables them to better understand the underlying concepts.

The practical work needs to be relevant to the project. To make practical observations of the human body is difficult, after all we are not advocating dissection! However, we do need to know something more about the body than pictures and even CD-ROMs can tell us. So we need to make as much use of our own bodies as possible. Eye and hair colour are obvious beginnings, and measurements of height, weight, hand spans, and foot size, are all useful data, and can be recorded in detail on the computer.

The class did much of this in their preliminary lessons, and recorded their results on the simple spreadsheets found in the *Number Magic* program already installed in their RM computers. They then decided to carry out simple experiments to test their individual reaction and pulse rates.

For the former they carried out the ruler catching test. One child holds a ruler vertically between the finger and thumb of their partner. Without warning the ruler is dropped and the second child must catch it as quickly as possible. The test is carried out a number of times, both to see if the catcher improves with practice as well as to find an average value for each child. Children read the measurement on the ruler, using it as an arbitrary figure for catching ability. They soon realise the importance of holding the ruler the same way up every time. They usually choose to have the zero centimetres at the bottom.

The class then arranged themselves so that there were two or three children per computer. As with so many primary schools nowadays, most of the computers are housed in a dedicated ICT room, so it was decided to use this as a class lesson. The children took turns both to hold and catch the ruler, and to fill in the data directly on to the spreadsheet, still using *Number Magic*. When this was completed, they used the formula builder to calculate an average figure for each child by placing the cursor in the cell that contains the formula, and then opening up the formula builder (activities menu) and clicking on ‘sum’. By highlighting the appropriate cells and by dividing by the number of cells, they obtained an average figure. Figure 4.4 shows a spreadsheet that was created.

The children were then able to explore the program’s potential for graphical presentation of their statistics. All the children worked quickly and enthusiastically. Even those with learning difficulties, and there were several, understood the practical work, and were able to use the computer. In many cases the level of computer skills were very high. As was the level of science, for several children were able to recognise genuine experimental errors, and correct or substitute them accordingly. The graph in Figure 4.5 shows each of the children’s individual reactions. The graph in Figure 4.6 shows an average reading for each of the children’s reactions. This was obtained by highlighting the name and average columns from the original spreadsheet. This is done by holding down the ‘CTRL’ key and clicking and dragging in the usual way.

At the time of writing, the class are researching their fitness by measuring their pulse rates before and after exercise. They step on then off a chair five or ten times in quick succession. They measure their pulse rates before starting (i.e. at rest), immediately afterwards, and then at regular intervals until it reaches the ‘at rest’ level. There is an obvious value in recording the results directly onto the computer program. With a judicial use of the proper graph, it will be easy to see

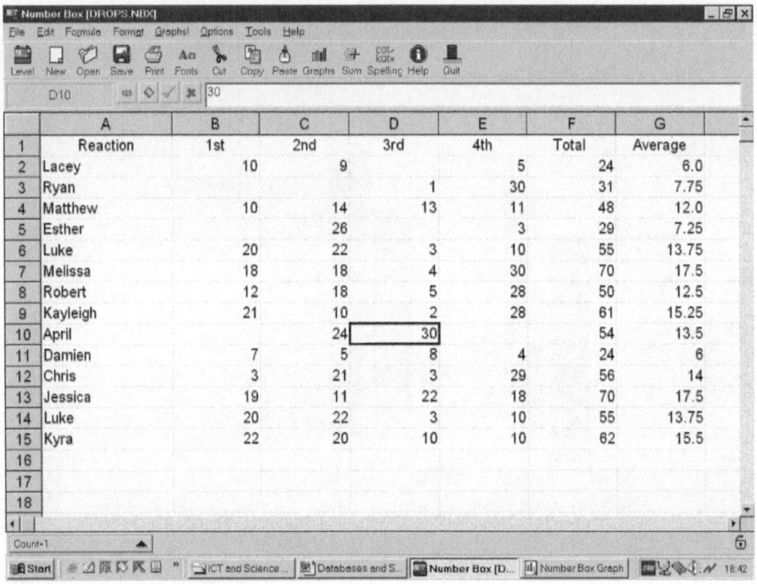

	A	B	C	D	E	F	G
1	Reaction	1st	2nd	3rd	4th	Total	Average
2	Lacey	10	9		5	24	6.0
3	Ryan			1	30	31	7.75
4	Matthew	10	14	13	11	48	12.0
5	Esther		26		3	29	7.25
6	Luke	20	22	3	10	55	13.75
7	Melissa	18	18	4	30	70	17.5
8	Robert	12	18	5	28	50	12.5
9	Kayleigh	21	10	2	28	61	15.25
10	April		24	30		54	13.5
11	Damien	7	5	8	4	24	6
12	Chris	3	21	3	29	56	14
13	Jessica	19	11	22	18	70	17.5
14	Luke	20	22	3	10	55	13.75
15	Kyra	22	20	10	10	62	15.5
16							
17							
18							

Figure 4.4 Spreadsheet used to record the number of ruler drops.

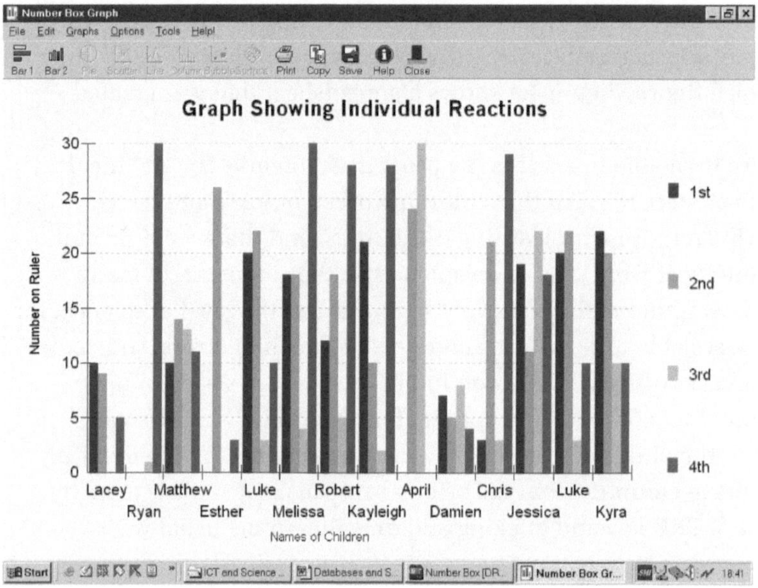

Figure 4.5 Graph showing each of the children's individual reactions.

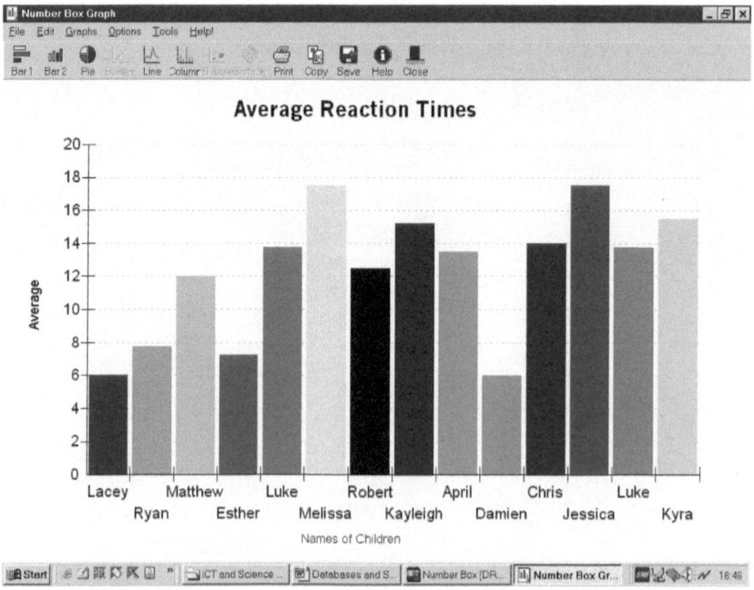

Figure 4.6 Graph showing an average reading for each of the children's reactions.

how long it takes for all or some of the children's pulse rates to return to 'normal'. The children will be able to analyse the data, and look for interesting patterns. They may, for instance, look to see if games players exhibit different results from other children. Obviously work like this needs to be carried out with sensitivity and understanding. It would certainly be done in such a way at Westbury.

Conclusion

Data handling is a fundamental part of all scientific enquiry. Without it there is little prospect of any genuine scientific investigation, or indeed scientific progress. As is frequently discussed throughout this book, the teaching of science in primary schools is an activity that revolves around first-hand scientific investigation and experimentation. The raw data that results from these investigations, and the subsequent transformation into information that can be presented and displayed in a variety of forms, be it numerical, alphabetical, alphanumerical, tabular or graphical, is correspondingly critical to the process. It is this data and information that gives meaning to whatever scientific concept or topic is being explored. It allows relationships to be identified, concepts to be explored and conclusions to be made. Through the use of databases and spreadsheets this can be done quickly and largely automatically, enabling the children to concentrate on the really important scientific concepts, the processes of understanding, enquiry and investigation.

Just like the two database studies, the children in this class were able to use 'real' results for their ICT. They first used the spreadsheet as a clear way to display the results of their work. This in itself would have been a valuable time-saving exercise. However, they soon found that the computer would enable them to take this further, and so were able to utilise this basic data in a variety of ways, examples of which are shown here.

Bibliography

Feasey, R. and Gallear, B., *Primary Science and Information and Communications Technology*, The Association for Science Education, Hatfield, 2001.

Chapter 5

Data logging

Data logging is the collection, collation and displaying of information gathered with the help of special sensors which are attached to a computer. These sensors could be specialised thermometers, light meters or even microphones. They could be used to measure the differing characteristics of any chosen environment, or the changing conditions during an on-going experiment. In the primary school this could be from an investigation in the classroom, or better still from an ecological project carried out in the field. Although a number of different types of sensor are available, it is sensors which log temperature, light and sound that are most commonly found within the primary school.

The place of data logging in the Primary Science Curriculum

The main advantage of using data logging as an integral part of primary science is that it provides a visual representation of environmental aspects that otherwise might be too difficult to see, feel, hear or accurately record, such as light, temperature and sound. It can log, record and display conditions that change slowly or rapidly over short or long periods of time. As humans we can tell when something is hot or cold, bright or dark or loud or quiet, but our senses can only measure these in very general terms and cannot provide an accurate value for them. We need special equipment, which usually means thermometers, light meters, or something even more sophisticated to measure sound or other phenomena. However, when using data loggers this information can be logged and displayed on the computer screen, usually as a line graph, giving immediate feedback of cause and effect (Porter and Harwood, 2000), which according to Feasey and Gallear (2001) is particularly important for younger children. For example, children can begin logging the ambient temperature, then dip the sensor into cold water and then into warm water and immediately see what effect this is having on the line graph that is being produced by the software on the screen. This is real information to which they have genuine ownership; they have started the logger, they can see what effect their actions have on the graph that is being drawn on the screen and they can see this immediately, in real time, as it happens. Events unfold before their very eyes! This is not an abstract graph produced from abstract data using traditional equipment; it is real, it is happening in front of them and it is highly accurate. This will help to enable the children to interpret the meaning of the graph, which in turn means that the main teaching and learning shifts from an emphasis on the collection and recording of the data towards the higher-order skills of actually interpreting what the graph is displaying. This is supported by the work of MacFarlane (1997), who through her research illustrated that quite young children can comprehend what a line graph is and the fact that it is showing one variable that can be constantly changing. She also demonstrated that this knowledge is transferable into other relevant contexts.

Developing higher order thinking

Too often children can get bogged down in what Harlen (2000) calls the 'drudgery', such as the need to take temperatures with traditional thermometers and record the data gathered by pencil-and-paper methods, and then laboriously produce graphs without giving any extensive thought to understanding what it is showing. We are not suggesting for one moment that this is not important; indeed, graphing and charting raw data into information is one of the key scientific skills, but there is no need to do this every time the teacher wants the class to take temperature readings. If ICT is not used, then this time-consuming 'manual' stage may have to be repeated every time the class is going to observe and record temperature readings. If data logging is used then this unnecessary repetition can be dispensed with; the time spent on the collection of the data can now be spent on the interpretation and analysis of the resulting information. As a direct consequence of this the children will be able to access higher levels of scientific knowledge and understanding. Harlen (2000) goes on to make the point that children often find reading a thermometer difficult and as a result findings may be inaccurate or they may forget to take an intended reading at a given time. Data logging can assist here in that it will be accurate and can be configured to take the log automatically. Because of this, these investigations can be reproduced accurately any number of times. This will ensure that investigations are always fair tests, a crucial aspect of primary science. As a means of additional support, many of the software packages that are used for sensing contain within them quite sophisticated analysis tools. These enable the user to interrogate the data that has been logged and displayed both accurately and in detail. This will enable the user to identify different conditions at different points during the logging period by placing a movable line and cross-hair on the graph. By placing the centre of these crossed lines over a given point, such as one of the dots on a scattergram, the exact value can be determined. An example of this might be where temperature or light and sound levels will be displayed at a given time. In this way data logging will help with the children's conceptual development because the more knowledge they gain of the processes required by the computer, the more they will understand the underlying scientific concepts needed. The results will be more accurate and therefore correspondingly more reliable, and as such the conclusions drawn will therefore be more relevant and meaningful.

An example data logging project

Although data logging was not available at the time it was written, a typical project where it could be used now was described by one of the authors in *Seasonal Science Series* (Williams, 1995). Here children were asked to compare various weather conditions within the school grounds. In this case the children were asked to take temperature readings in various places, in order to try to discover if there were any wind-chill effects evident. All the readings would then have had to be recorded carefully by hand, and the conclusions and recording written up later. Whilst some of this basic recording could have been done with the help of a computer, the actual measurements would have had to be carried out 'by hand', and only when the children were present.

Sensing the weather

For the more conventional study of local weather conditions, there are now complete computerised weather stations that are available to primary schools, and we need to consider these here, as they incorporate the processes of data logging. These consist of measuring and recording equipment – effectively data loggers – fixed to a pole outside and positioned so that the meteorological elements will not be obstructed or influenced by local factors such as walls or trees. This invariably means that the equipment is fixed in a high position such as the roof of the school hall. The measuring equipment usually consists of a cup anemometer to measure wind speed, a weather vane to measure wind direction, an electronic version of a measuring gauge to measure rainfall, a pressure sensor to measure barometric pressure and a temperature sensor to measure the temperature. Additional sensors can also measure other environmental features such as humidity and light levels. The measuring equipment is connected to one dedicated computer within the school, often positioned in a shared area so that it can be seen by as many people as possible.

Although the computer can be used for other purposes whilst it continues to log, it is primarily used for continuously and constantly displaying the weather conditions in the form of graphs and charts. This information can then be interrogated and analysed either in real time as the data is collected, or can be rearranged into different graphs for further interrogation and deeper analysis at a later time. Also, the children do not need to go outside to collect their weather data. This is done from indoors, a very important consideration if it is winter or if the weather is bad. Where there is a need to collect the data on a daily or twice-daily basis as part of an ongoing project or experiment in order to maintain a fair test, then the computer will do this if so programmed. A major benefit of this system, or indeed any data logging, is that it ensures that the data that is collected is accurate, which can be a big problem when collecting data of this nature. For example, thermometers can be difficult to read, and wind speed can be difficult to measure as the equipment can be inaccurate and difficult to use properly, especially if the children are wearing gloves or have their hands full of other associated bits and pieces such as clipboards and pencils.

It can also be used to create and display other weather information that in the past would have needed a separate instrument to measure and record, such as a maximum and minimum thermometer. Although this would have recorded the daily maximum and minimum temperatures, the thermometer would have had to be reset each day with a magnet, with the current air temperature being measured on a separate thermometer. When an electronic weather station is used this can be done by one temperature sensor, with the times of the highest and lowest temperatures being extrapolated from the collected data along with the times at which they were measured by the software rather than the hardware, as would have been the case in the past.

Whilst there may be at first glance many advantages in using this ready-made, automatic set of equipment, it needs to be considered that there is no practical aspect involved. It could well be argued that this represents a significant disadvantage as far as good primary science is concerned. As the weather conditions are logged continuously and automatically by the hardware and software, the children will not be collecting their own information. Indeed, they will probably not even be starting and stopping the logging process, putting the sensors outside and measuring the various environmental conditions. All this has already been done for them, so there is little for the children to do but observe. If the computer is positioned away from a window they may even have little or no idea about the state of the weather outside. They will be unable to relate and connect the information that is being displayed on the screen in front of them to what the actual weather conditions are. Without practical involvement, this information may be too abstract, as the graphs can be difficult for younger children to read and interpret, and it can be difficult to draw conclusions or to identify relationships between one set of data, such as temperature and another, such as humidity or barometric pressure.

An ideal solution, and one that could be used in the *Seasonal Science* example, is a compromise between the completely automatic example of the weather station described above, and using computers to record data collected 'by hand'. This can be done through the use of simple equipment that will collect electronically all the data independently from the computer. Once this has been completed, it can be connected to a computer for the downloading and display of that data. One such piece of equipment is the EcoLog system (Figures 5.1 to 5.3) produced by Data Harvest, consisting of a small interface box, leads, photocopiable Curriculum Notes, and its software – *Sensing Science.* There are three analogue sensors for measuring the air temperature, light and sound levels. EcoLog Plus allows for two more sensors to measure humidity and barometric pressure. The software allows all the information to be displayed in a variety of ways – as a change of colour, numerically, graphically, or on a dial. The latter shows the information collected over one hour. It offers a line graph, which allows children to draw predictions on the screen, and then compare these with the actual data. It can also take instant readings, which are stored on a spreadsheet for later analysis. Here the children are involved at every step; they can collect the information and they will be able to connect what they have seen with the data that they have gathered.

Data-logging activities

Some sensors can also be used as inputs for control boxes, as explained in Chapter 6, 'Control technology'. They sense changes in environmental conditions, such as movement, air temperature or light levels, and the computer is then programmed to create a linked and corresponding response from the outputs. These could include opening or closing a window, or a curtain on a model. Although this is discussed in greater depth in the Chapter 6, it serves here to illustrate the flexibility and the range to which sensors and data logging can be put. It also replicates a range of real-world situations such as automatically opening and closing doors and windows in modern buildings when there is a change in temperature or light intensity.

It is not difficult to imagine the many science projects in which such data-logging equipment would be invaluable. There are many such examples listed in various texts, including Data Harvest's own manual, which contains some excellent activities for children right across the primary phase.

Figure 5.1 EcoLog from Data Harvest.

Some potential classroom investigations might include checking the effectiveness of sunglasses or the reflective properties of materials by using light sensors. Here a light source such as a torch can be shone through different types of sunglass lens to see which is the best at filtering out light, or the light can be shone onto a range of materials to see which one reflects the most or least. The opaqueness of glass could also be measured, as well as the blackout properties of different materials. A range of materials could be provided – which material would make the best pair of curtains? Temperature sensors are ideal to use in the now standard insulation experiments. For example, four different types of cup could be provided containing warm water filled to a similar depth. These are then covered with similar lids to ensure a fair test, and then the temperature sensors will then be used to log the changes in temperature over a given period of time. Which cup holds the temperature for the longest, and thus has the best insulating properties? Similar cups could have a range of materials wrapped around them, and again, which material has the best insulating properties? We suggest that as a variation to this standard experiment, children should use a variety of bird feathers as the insulating material. This can be part of any bird

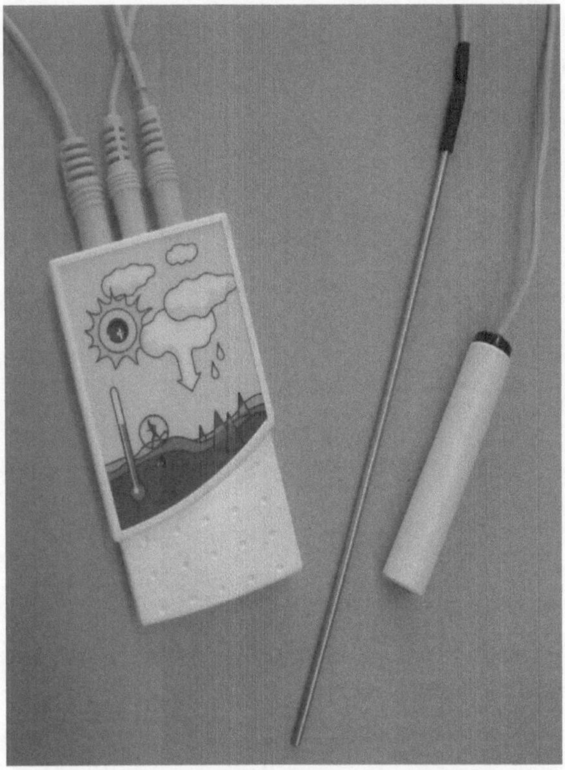

Figure 5.2 EcoLog with temperature and light sensors.

study, or any general study of living things, as well as the study of properties of materials. Whatever the aspect of science being studied, teachers should find opportunities to use a range of non-standard materials that could be, and in many parts of the world are, used for insulating. Mud, straw, leaves, as well as reflecting materials such as metallic foil, could all be used. Of course, one cup must remain uninsulated as a control for the rest of the experiment.

Another topic, which we think is of special interest, is to measure the actual temperature, light intensity and noise level simultaneously in the classroom itself, during the course of the normal school day. As discussed above, the feedback will be direct and immediate in that all the readings will relate directly to the children themselves. The noise is theirs, and they will have been working within and been affected by the temperatures and different intensities of the light. When analysing the data the children can see if any of the three factors have any connection. For example, does the temperature or light intensity affect the noise level, and if so why? This is a particularly good topic for Key Stage 1 children, not we hasten to say because of the noise levels, but because it does

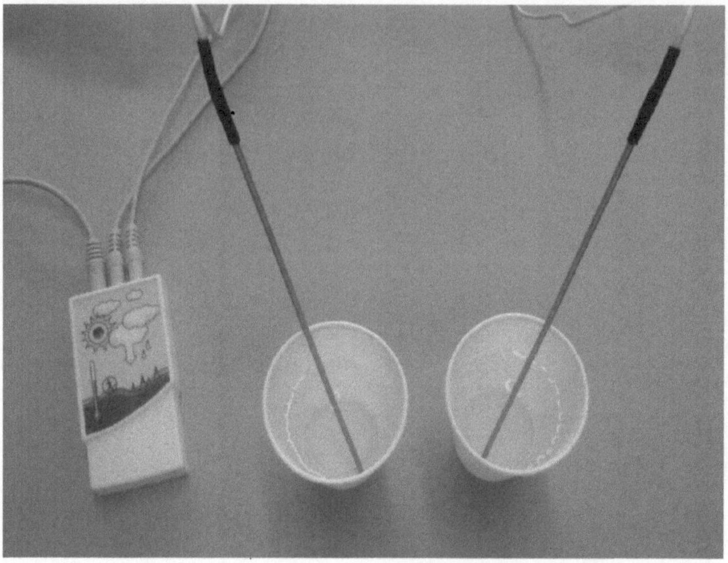

Figure 5.3 EcoLog: Two temperature sensors logging warm and cold water.

have a direct relevance to the children themselves, and because the software can give instant feedback and immediate statistical results.

Case study

Although we have suggested several classroom activities that will make interesting topics, it is surely in the field, away from the classroom, that data logging comes into its own. The authors remember various ecological field studies in which children made many observations of the environment, but had to make the measurements by estimation and comparison only. Two such examples were a study of a pond, and a comparative study of an oak tree with a Scots pine, which a Year 6 class from Elaine Junior School, Strood, undertook during a field trip to the Weald of Kent. Although only two trees were studied, the amount of information gathered was legion. Insect and other invertebrate life, evidence of mammals and birds and where in the trees they were found, fungi and other parasites, the shapes of the leaves, fruit and seed dispersal, the different barks and even the algae on the trunk were all studied. When the information was collated, it was possible to construct a detailed picture of the trees as two very different living ecological units, and even construct food webs within and around each of them.

All this was done before computers found their way into schools. Imagine how they could be used in these topics if they were to be carried out today. In the pond study the temperature of the water was taken by lowering thermometers into the water at various depths. A sensor, as long as it was made waterproof, could do this very easily. A light sensor could be used to see how far light penetrates the water, and whether this has any effect on the temperature. Who knows what a sound sensor might reveal! However the scope for ICT in the second of the projects is greater still. Databases and spreadsheets could be used for the correlation and identification of animals, clip-art software for the different shapes of leaf, tree shapes, and perhaps the design of the food webs. Data logging with attached sensors would be used for the measurement of temperature, and for the light intensity, either within the canopy of the trees or on their trunks when measuring the density and spread of algae or lichens. The sound sensors could even be used to measure the dawn chorus! To complete the picture, the whole project could be made into a video presentation.

Thanks to Eric Parkinson, now at Christchurch College for his recollections of the pond study when he too was a teacher at Elaine Junior School.

Figure 5.4 shows an impression of what the graph of the temperatures and sound levels from inside one of the trees, mentioned above, might look like. It does not show the complete window, and the columns on the left would list details of the time, temperature and a measurement of the sound. There might also be a fourth column for comments, a useful addition to allow added detail, to what was, for the temperature at least, an accurate reflection of a bright day in late March.

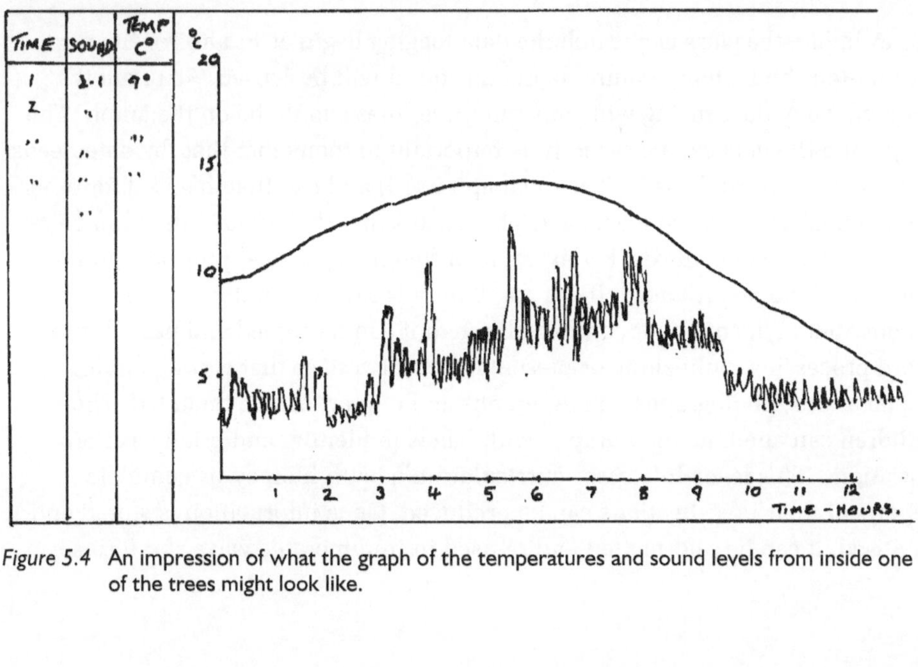

Figure 5.4 An impression of what the graph of the temperatures and sound levels from inside one of the trees might look like.

The power and flexibility that data logging offers the teacher clearly has significant implications in terms of teaching and learning, classroom management, organisation, obtaining the necessary resources, and in health and safety. Unlike most other areas of ICT, data logging can be started and then left to record automatically, so the children and teacher can go and do something else, either related or unrelated to the data-logging activity. It can be set up to record data over considerable periods of time – perhaps up to a week, so does not need constant attention. Assuming that the data logger is being used when directly connected to a host computer, once data logging has started then it will continue to collect data automatically until it is manually stopped or the programmed logging period has been completed. As far as organising and managing data logging is concerned, this may mean that the teacher can give a class explanation, start the log and then not revisit it again until the following day or even the following week, except perhaps for the occasional check to ensure that it is still working properly. It would be advisable to take time to briefly examine the resulting graph periodically, so as to keep the children interested in what is after all a practical topic, as well as to try to predict what will happen next. It is also important that the children have an opportunity to have 'free play' with the equipment first. This will enable them to understand how the sensors work, how they are calibrated and how the line will be graphed against a given change in condition or value. It will also ensure that when they have the opportunity to devise their own future investigations they will have the knowledge and understanding to make key decisions such as which of the sensors are appropriate for a given purpose and for how long and where to take the log.

For example, the class can watch the data logging begin at 11 am. For the next twenty-four hours temperature, light and sound will be sensed. At 11 am the following day the sensing will stop and three lines should be on the graph. The graph and the data can be saved (it is important to remember that the data needs to be saved separately as well as the graph itself) and can then be used for other applications such as a spreadsheet. This means that this original data can be re-used in a range of contexts. For example, different types of appropriate graphs and charts can be produced. If the graph only is saved then it cannot be reconstructed in any other context, although it can be copied and pasted into word processing, publishing or presentation software so that accompanying explanations, photographs, tables, graphs and clip-art can be included. The children can annotate their graphs with labels to identify and briefly explain variations. This leads into cross-curricular links with literacy, as complete reports of their investigations can be produced. Once information is stored and digitised, it can be reformatted and re-used in innumerable ways and forms.

Although arguably not directly data logging as such, another related activity involves the use of electronic scales. These are very sensitive weighing scales, supplied with software and cabling which enables them to be connected to a computer. This can be configured to log the data provided by anything that it is weighing after every few seconds, or indeed over any given timescale. The resulting weights are then automatically pasted into a spreadsheet every few seconds, and from this point the information can then be used for a range of scientific projects. This apparatus would be particularly useful for longer-term data logging projects, where very slow changes in state from a liquid or solid to a gas can be recorded. This could be detecting the transpiration rate from plants, where the plants are watered and are then left on the scales. As the plant draws the water in through its roots and then transpires the moisture as vapour, the total weight of the plant will correspondingly decrease, as the water content in its container is utilised. As the plant loses water, the leaves will wilt, and it is during this period that the maximum weight loss will be recorded. This will be detected by the scales and will be recorded in the spreadsheet. When the data is graphed, the gradual reduction in weight will be clearly visible and can then be analysed. In a similar way evaporation can be detected; a saline solution can be placed on the scales and as evaporation occurs, leaving salt crystals behind, the scales will measure the changes and the spreadsheet will record the corresponding weights. Again, this is a very good way of ensuring that something that otherwise would be very difficult to detect with the naked eye, simply because the process is so slow, can have a visual representation. For younger children, simply placing an ice cube in a glass on the scale pan, and seeing if there is any difference in weight during and after it melts (and itdoesn't matter if there is not!), can be very interesting. We have often studied how children feel about the conservation of volume, so how about the 'conservation of weight'? Does the weight change, and if so why? The teacher needs to make sure that records are taken at regular intervals, as there will come a time when the melted water will begin to evaporate. Michael Faraday, who studied many things which we now recognise as essential components of the primary science curriculum, including how melting ice behaves, would have loved one of these tools!

Extending children's thinking through focused questioning

It is very important to ensure that along with other areas of primary science, the place and role of questioning is paramount. The teacher needs to ask questions that are focused, detailed and searching, yet are sufficiently open-ended to ensure that the children can discuss their ideas and their findings. It is insufficient for the children to simply be asked to describe a graph; although a description clearly is needed, there should also be some analysis involved. The children need to explain in their own words and with their own ideas why something has happened. Although their understanding of scientific concepts and ideas may need some development, this reasoning and hypothesising is crucial. This is the point at which in-depth teaching and learning is taking place. It is by extending children's thinking in this way that the necessary depth and the value-added component of using ICT to support the activity comes into its own. Thinking back to our twenty-four-hour data-logging activity mentioned above, the following open-ended questions could be asked:

- Describe the lines on the graph. What are they showing? Why do you think this has happened?
- Why do you think that there are occasional 'blips' of light when it is dark? (For example car headlights will show up on a data log.)
- Is there a difference between the sound intensity during the night as opposed to during the day?
- Is there a connection between sound intensity and the fluctuations in the temperature during the night?
- Do you think that the readings reflect an urban or a rural environment?

The teacher could then ask more closed supplementary questions that take a more interpretative focus about the actual graph itself:

- At what time did it start to get dark?
- How long was it dark for?
- When did it start to get light?
- How long did it take to get light?
- Was it a cold night?
- When was it at its coldest?

Whatever approach is taken, it is important that the teacher asks the 'right' questions. Along with the important primary science skills of hypothesis, prediction, observation, recording (including where appropriate a well-considered conclusion), questioning is a key element of data logging that needs to be carefully considered at the planning stage. Indeed, such is the potential for data logging to deliver good primary science, it is the only type of ICT application that is specifically mentioned in the National Curriculum for England (2000), being detailed in both the Science (Key Stage 2, Sc1, 2f) and ICT documents (Key Stage 2, Developing Ideas and Making Things Happen, b).

Health and safety

As in all aspects of science, so in ICT, health and safety considerations should be an integral part of the planning and you should consult your school's Health and Safety Policy. For data logging, one advantage of using sensors is that they are both easier to use and are more robust than more traditional means of recording environmental conditions such as a glass thermometer. When deciding what to log, it is important to remember that good primary school science is a practical activity in which children can engage in complete safety. Therefore it is important that only safe materials and conditions are used. Obviously great care should be taken when using hot water (not boiling) or ice, and therefore an adult should be on hand to do any pouring of liquids. Where possible it is better to use water at room temperature, perhaps containing ice cubes to help to lower the temperature. Sensors will not detect the temperature of solid ice as accurately as cold water. Using cold and warm water will still give a sufficient range in temperature to give a good reading and a correspondingly meaningful line on a graph, but will be completely safe and will allow the children to handle all of the materials used in the investigation.

Another factor that should be made clear to the children is that the materials used are for experimentation and are not there to be consumed. That they must not eat or drink any food or liquids being used as part of their investigations, although obvious to us, needs to be emphasised at an early age. Quite apart from the fact that they have been handling foodstuffs and placing sensors into them that might be contaminated, it is good practice to ensure that young children learn to carry out experiments in a safe and correct way. Later on during their school careers they will be handling substances that may be far more toxic than warm water.

An additional and highly relevant factor in the safe conduct of these experiments is that children need to be reminded about taking care when using liquids near computers. Liquids can damage a keyboard or even worse, run down a network or mains power cable into a floor socket or network point. With data logging, using both computers and liquids together cannot always be easily avoided. However, the interfaces and sensors often have long cables, thus ensuring that substances do not have to be used near computers. EcoLog has the added advantage in that it can be used whilst disconnected from the computer. As long as the children are made aware of the potential risks and the teacher undertakes a risk assessment prior to the commencement of the activity then there should be no problems. Although health and safety considerations are very important, they should not prevent an activity that can add a great deal to the teaching and learning of science.

Conclusion

Data logging is potentially one of the most powerful aspects of primary ICT in science. The ability that it offers in allowing the child to 'see' temperature, light and sound unfolding on a graph on a computer screen is one that should not be underestimated, and is one that can genuinely develop a range of scientific concepts.

Bibliography

ASE, *Be Safe*! third edition, Association for Science Education, Hatfield, 2001.

DfEE, *The National Curriculum for England*, Department for Education and Employment, London, 1999.

Feasey, R. and Gallear, B., *Primary Science and Information and Communication Technology*, The Association for Science Association, Hatfield, 2001.

Harlen, W., *Teaching, Learning and Assessing Science 5–13*, third edition, Paul Chapman Publishing, London 2000.

Lambert, M. and Williams, J., *Project Ecology, Animal Ecology*, Wayland, Brighton, 1987.

MacFarlane, A., *Information Technology and Authentic Learning – Realising the Potential of the Computer in the Primary Classroom*, Routledge, London, 1997

Porter, J. and Harwood, P., *Datalogging and Datahandling in Primary Science*, MAPE Focus on Science, MAPE, Newman College, 2000.

Williams, J., *Seasonal Science Projects, Winter Science Projects*, Evans Brothers, London, 1995.

Chapter 6

Control technology

When we see television pictures of robots assembling and painting an assembly line of cars, we are viewing an example of control technology in action. A simpler example, and one that we meet many times in our lives is the sliding-door mechanism in the local supermarket. These doors open as we approach them, and the sliding doors are activated by a signal from us. It could be via a light-sensitive electric cell, or by pressure exerted by our bodies on a pad beneath our feet. Whichever example we chose, some form of computing is involved.

The nature of technology

We should at this stage make it clear that to us 'technology' is akin to science. It isn't its servant or in any way a lesser subject, but it is historically connected to science. In some schools, both primary and secondary, it has recently become part of the art syllabus, although recent changes in the National Curriculum do seem to have shifted the balance back to a scientific or an engineering approach.

Before we examine the role of the computer in control technology, we should first look at how this process has evolved both as a subject within the curriculum, and in the way that it has been taught. In the primary school at least, technology, and particularly control technology, was taught – if at all – as part of the science curriculum. This was before the advent of the National Curriculum, although it is interesting to note that in the very first National Curriculum Science document, ICT had a section of its own. This was linked with Microelectronics, a very technical aspect of Science. It was not until later that control technology officially became part of Design and Technology, a completely separate subject. However, if we are to consider technology as a close relation of science, then control technology, with its strong ICT input, must be part of this book.

It might be appropriate that to teach the mechanical and electrical components of Design and Technology (the science part) the learning should approximately follow the same pattern of development as the subject matter itself. This learning would therefore begin with young children making very simple models that nevertheless have a simple mechanical working part, such as a pop-up Christmas card. Static models with no working parts can rarely be said to be technological, although they may use technological processes in their making. Of course, the technology needs to be aesthetically pleasing, but must follow scientific principles or it just won't work. It is possible that the recently completed Millennium Bridge, spanning the Thames in London might be one such example. Although this elegant structure seems to be a form of suspension bridge, and that the technology of these bridges has been understood for many years, what apparent design fault caused it to sway when people first walked across it?

When we first teach children about simple mechanical processes, we often ask them to make models out of card and later wood. These models will have simple mechanisms such as levers, sliders or wheels. The working parts may be held together by pins or paper fasteners, or wooden and metal axles. However they are made, they will be powered by hand, or perhaps if the model is a watermill or windmill or a land yacht, then by wind or water. Since the children will also have had experience of basic electricity in their work with bulbs and batteries, there may come a time when they will want to harness electricity to drive the model, and to enhance it by including lights or buzzers. This, for many primary school children, is often the final stage.

What is control technology?

Children will need to be present to switch on many of these models. However, with control technology they need not be. Models of level-crossing barriers, traffic lights, or lighthouses, can be made and worked by a previously programmed computer. In life, all these are seldom, if ever directly controlled by hand. When we first started teaching Design and Technology, there were about ten manned lighthouses in the UK, but today there are none, as all are now controlled by computer, as is almost every level crossing.

There are two basic approaches to the teaching of control technology. We can either have the children make the model to be controlled, and then learn to write their own programs to control it, or we can buy a kit. Kits can be bought that provide easily assembled models, driven by computer. These programs exhibit a range of simple clear icons, and all that is needed is to click on them to make your model work. The computer is not just an on/off switch (although it can always be used in this way), but the icons, used in the appropriate order, will allow the model to work to a pre-ordained pattern.

One such program is the *Robolab* software, often pre-installed in primary school computers. This can be used with the *Lego Mindstorms* equipment. This has a separate Lego brick that contains the control box for the computer. It is an entirely separate unit, has its own inputs and outputs, and has wheels so that it can be used as a self-contained robot. It is not wired to the computer like a static control box, but is worked by infra-red signals.

The programs are organised in easy steps. Pilot one, for example introduces the icons to work the motors, and introduces the directional commands. Pilot two shows how motors can be used for steering by changing their speeds, as well as further commands for direction and timing. By Pilot three, children can use a sequence of commands to change the robot's direction, so that it can follow a planned pathway, and use other icons, such as ones that allow for continuous and repeated commands. There are subsequent programs which children can use for further and more advanced study.

This is a more simple approach and less time consuming than building and programming your own model. Children take to it very quickly and obviously enjoy the work. Its costs are comparable with other control systems, particularly if the school already has the software installed. However, we do need to think carefully about what we want to teach. There is little science involved in this approach, and indeed very little constructional technology. Simply clicking onto an icon does not in itself help the children to understand the necessary commands needed to make the model work. The children need to be asked what they think the icons mean, and what words could be used to describe them. This would help them to design the correct sequence of icons, and so build up a more useful program. Finally, if something goes wrong with the pre-designed kit model, can children even discover where the fault is, let alone fix it? Because children actually build their own model, the alternative, if slower, method will not only require them to plan and write programs, but enable them to check, and where necessary replace each separate component. Each step will help reinforce previous knowledge and experience.

For instance, let us suppose the children have built their own lighthouse. They will have tested the light by connecting it via a simple switch to a battery. If they were not sure how to construct an electric circuit, they certainly will have found out by the time they had completed the model. They could now use a computer to control the light, for at this stage the teacher would be asking the children to think how a computer could help. It goes without saying that as in all projects, the teacher would have presented the children with plenty of background information. They would have been introduced to the appropriate books, and may even have seen a suitable video, or in this case visited a museum, or had a talk from someone from Trinity House. They may also have had previous experience of some of the equipment during other lessons or topics. Light-sensitive cells are not difficult to understand and could be a part of a simple model, perhaps, made by some Year 4 children.

Control technology requires 'inputs' and 'outputs', and for these to work the computer will need a control box, together with the necessary leads and wiring. The control box, connected to the computer, passes on the processed information to the model. On a standard control box there will be connections for the various lights, motors or buzzers that work the model. All these are collectively called the 'outputs'. This is illustrated in Figure 6.1.

On the left side of the control box are the inputs, on the right side are the outputs. The wiring connects to the model. When a pre-programmed condition activates the input, this causes a response from the outputs which illuminates lights, sounds buzzers, etc. This model was designed and made by an Initial Teacher Primary Trainee.

In the case of the lighthouse there may be only one output, the bulb. This will be programmed to come on at a certain time, flash two or three times (every lighthouse has its own signal), and then switch off. This sequence has to be repeated for as long as necessary. The children should be asked how this sequence could begin. They might suggest that they start it by pressing the return key on the computer. It should be explained to them that using the computer as a switch in this way defeats the object of control technology, in that someone has to be present to press the key. They might then suggest a time-switch or better still a light-sensitive cell, which turns the sequence on or off when it gets dark or light. Which way round is a matter of common sense. This, just like the switch activated by a customer approaching the supermarket door, is the input.

Perhaps because of the time factor, or cost, or simply because its importance is not realised, very little control technology seems to be taught in primary schools at the moment. It may be that control technology is a casualty of the overcrowded curriculum, or simply that teachers do not have the necessary expertise. Perhaps it is a combination of all these reasons. It also seems that

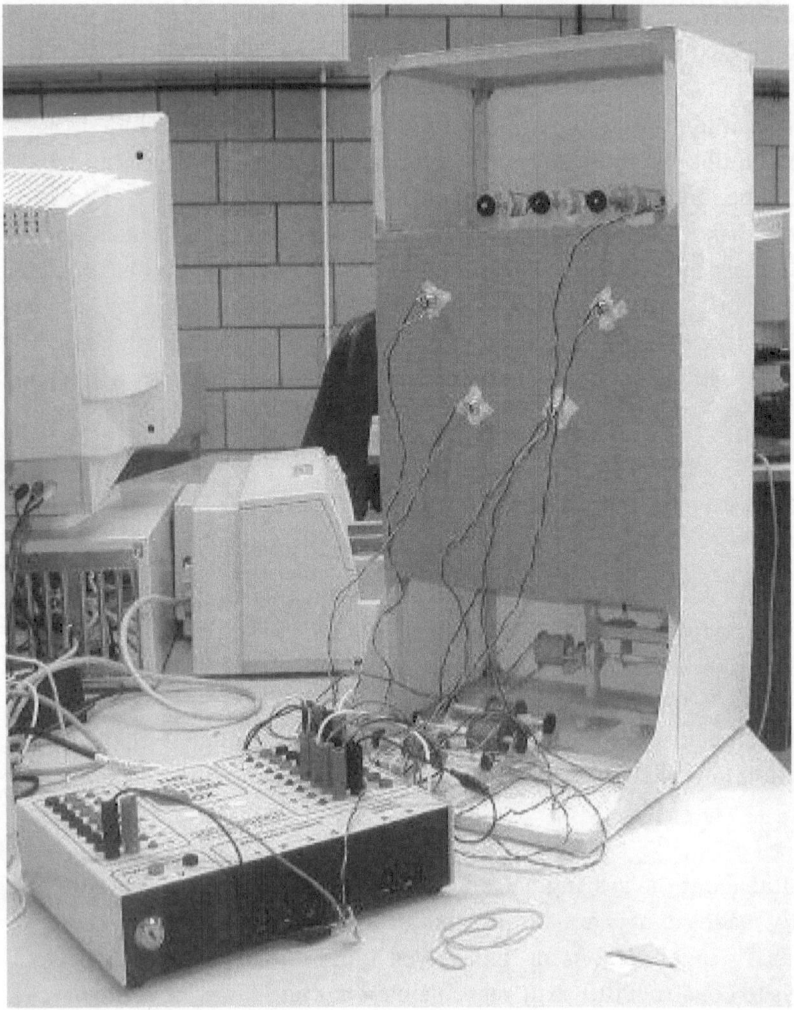

Figure 6.1 Control box and model.

crossing subject boundaries has gone out of fashion. This is a pity, for by using an integrated subject approach to learning, children are given the opportunity to realise that knowledge in its widest sense is seldom parcelled up into little packages, marked English, Mathematics or Science. To design and build a model, link it to a computer and program it, includes elements of many subjects. Many aspects of the National Curriculum requirements for these subjects can be easily met through this cross-curricular approach, and control technology can provide an apposite vehicle.

Let us use the lighthouse example to find out how we might do this. We would have first designed the lighthouse, which can involve historical and geographical research. This could be designed on a computer, but pencil, ruler and paper will do just as well at this stage.

Let us imagine that to make it more interesting, the children have included, as well as the light bulb, two more 'outputs'. A buzzer (to act as a foghorn) and an electric motor. The motor will be mounted at the top of the structure and will have a coloured disk fixed to the spindle as another kind of signal. Motors are often a part of many models such as lifts, cranes, or drawbridges, and we need to show how they work. Children will soon discover that electric motors need to be geared down to make them work at reasonable speeds, or to lift heavy loads. They will have already used motors to drive buggies or to lift loads, so will have had to face such problems before.

Using questioning to access higher levels of understanding

The teacher may need to remind them of this, by asking questions such as:

- 'What will happen if we try to lift this very light load with the motor?'
- 'What will happen if we try to lift a very heavy load with this motor?'

If the children had had some recent experience with this equipment, then the questions would more likely be ones of the 'Do you remember when we ...?' variety.

When a decision about the gearing has been made, the teacher can choose to fit a commercially made set of gears, but it is better if the children make up there own gear trains. By making them for themselves, children will not only learn a variety of ways to construct different gear mechanisms and know how they work, but will begin to understand the principles of gear ratios, and the concept of torque.

Constructing gear trains

You will need:

- 2 lengths of 10 mm squared section wood, each 26 cm long
- 2 lengths of 10 mm squared section wood, each 12 cm long
- 3 plastic cotton reels
- Three 14 cm lengths of dowel rod. Each rod should pass through the centre of the cotton reel and be as tight a fit as possible
- Thick card
- PVA glue

- Rubber bands
- Masking tape
- Electric motor with plastic spindle and spring mount.

You need to measure out three points on the long sections of wood. There needs to be one at 8 cm, one at 14 cm, and the final one at 20 cm. Then drill holes through both of the long lengths at this point, so that when the dowel is put through the hole it will run freely. Carefully glue the lengths of wood together to make a frame, ensuring that the short lengths are both glued to the inside of the long lengths. Strengthen each corner by attaching a cardboard triangle with PVA glue and then wait until it is dry.

Figure 6.2 shows the side view (not to scale) of gear train frame with triangular-shaped axle bearings as shown in the photographs. This dispenses with the need to drill holes in the frame and enables the frame to be used for other purposes depending on the level of work required. When cardboard wheels are attached it can become a chassis, or without any attachments it can become a picture frame.

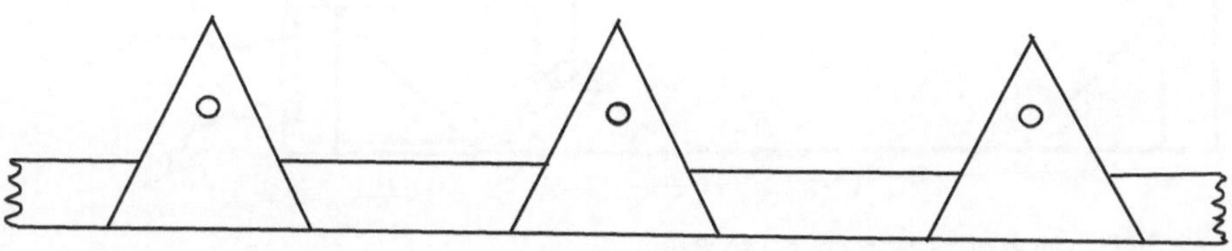

Figure 6.2 Side view (not to scale) of a gear train frame.

Pass the dowel rod through one side of the frame and fix a cotton reel and a selection of rubber bands onto each dowel. Once you have passed the dowel rod through the other side of the frame, you will need to make sure that the cotton reel is securely fixed to the dowel. You might need to stick a layer or two of masking tape onto the dowel so that the cotton reel is a tight fit. The cotton reel **must** be an integral part of the rod. Offset the centre cotton reel, so that they are not all one behind the other. Fix the motor holder to one end of the frame, and when secure, put the motor in place. Now connect each dowel and cotton reel fixture with the rubber bands so that one end of the band is round the cotton reel and the other round the dowel. The end of the final band goes round the motor spindle.

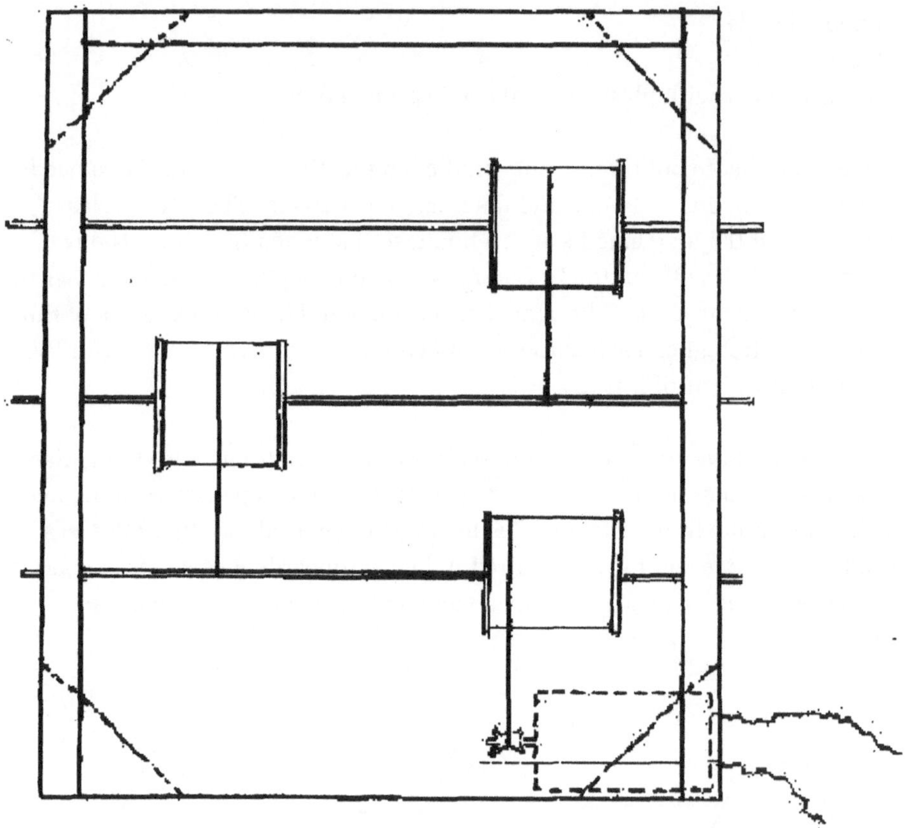

Figure 6.3 Plan view of a gear train (not to scale).

Figure 6.3 shows a plan view of a gear train (not to scale). Note the position of the motor with wires to the power source, and the card corners on the underside of the frame. The motor can be positioned at either end, as long as the rubber bands are adjusted accordingly.

When the motor is connected to a battery, you will notice that each cotton reel rotates more slowly than its neighbour. This is simple a matter of mathematical ratios and can be calculated. The principle is no different from the cogwheel gears on a bicycle, but it is easier to make in a primary school. We do use cogs in primary science and technology, but usually only to change the direction of movement. If a length of cotton is fixed to the slowest rotating cotton reel, the other end can be tied to a drawbridge, level-crossing barrier, or used simply to lift a sensible weight. In each case the object will be moved slowly and easily.

Figure 6.4 shows a gear train, and Figures 6.5 to 6.9 show gear trains in use.

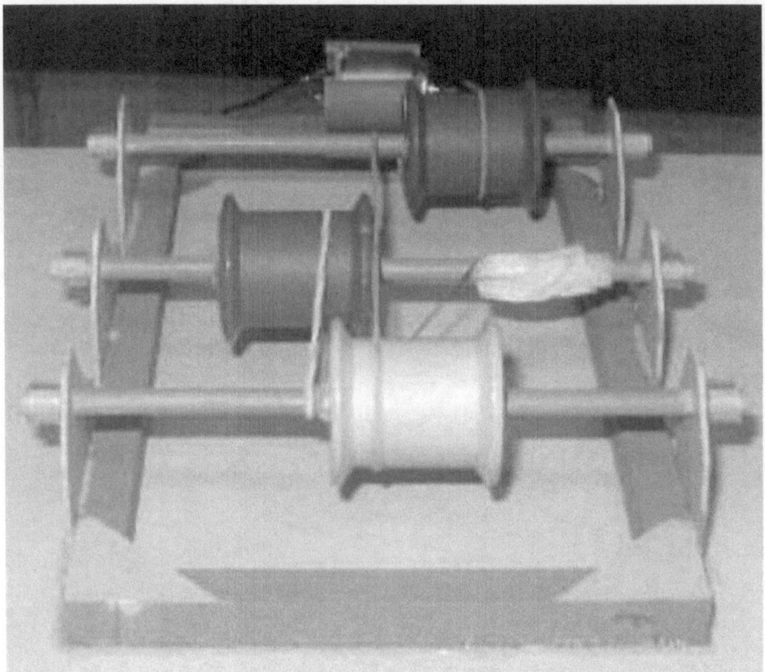

Figure 6.4 A gear train.

Figure 6.5 A gear train *in situ*.

Figure 6.6 Close up of a gear train *in situ.*

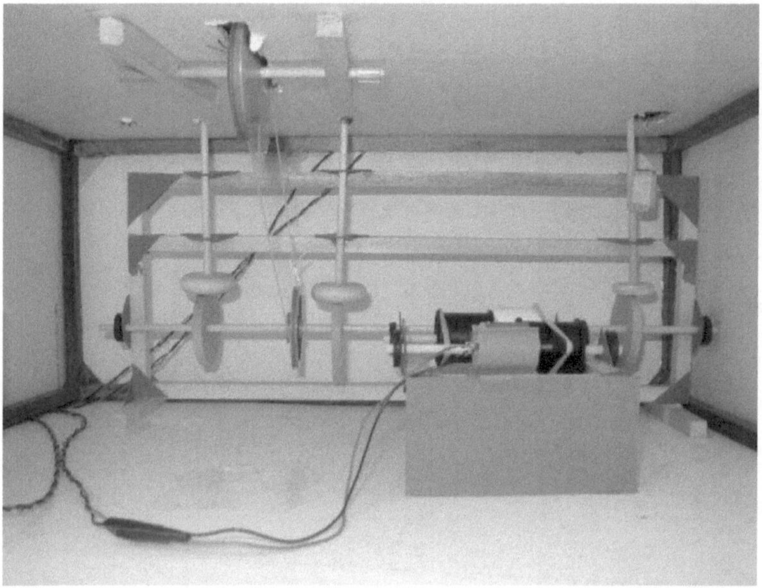

Figure 6.7 The gear train is connected to a set of cams which are used for alternating the movement of the sculptures fixed to the top of the pistons.

Figure 6.7 shows a model in which the gear train is connected to a set of cams which are used for alternating the movement of the sculptures fixed to the top of the pistons. These sculptures are not shown as they are above the false ceiling of the model. The gear train fixed to the cams in the model shown in Figure 6.8 allows for the cat's mouth to open and close. A close up of cams attached to a gear train is shown in Figure 6.9

Motors can rotate in two directions, clockwise and anti-clockwise, or in control terms, backwards and forwards. Most control boxes have about eight sets of terminals for each of the 'inputs' and 'outputs'. A motor will need two, so four of the 'output' terminals will be required, but, for this model, only one 'input'

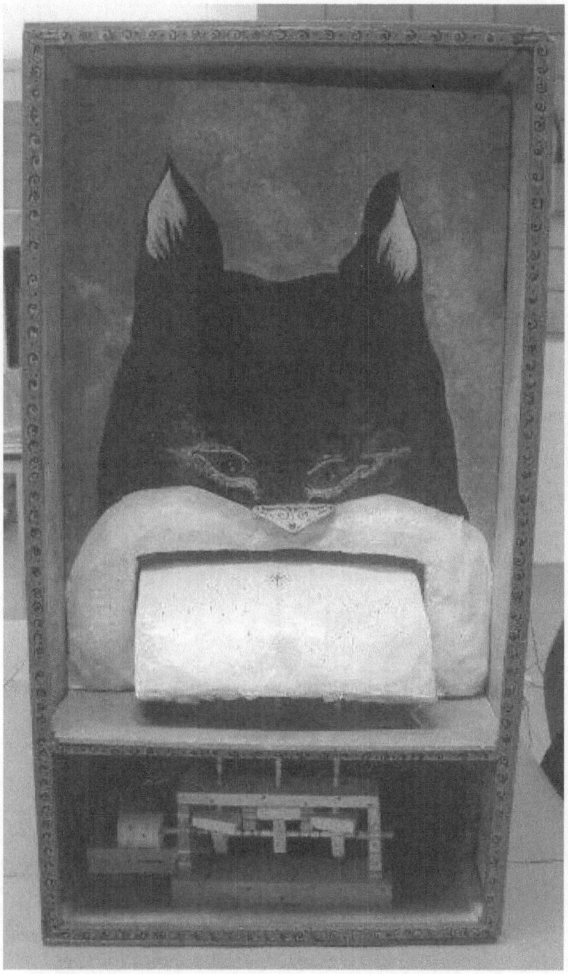

Figure 6.8 The gear train fixed to the cams allows for the cat's mouth to open and close.

terminal. Each pair of terminals is numbered, so the input light-sensitive switch will therefore be labelled as 'input one'. Before connecting to the control box, all the 'outputs' need to be tested individually with a battery.

Once the model is connected to the control box, the children can begin to work out their program, for by this time they will know what their model can do. They should first write the program down on paper as a sequence of events, or even as a flow chart. They will then have a clear idea of what they require their model to do. They might write the following:

Figure 6.9 Close up of cams attached to a gear train.

When it gets dark, the computer turns the light on for three seconds, then off for two, then on again for three then off again for five seconds. After these five seconds the buzzer sounds for three seconds and the motor turns forwards for three seconds; both at the same time. All stops for a further five seconds, and then the sequence is repeated until it is light.

(In order to simulate darkness, all that is needed is to cover the light sensitive 'input' with a finger.)

To turn this into language that the computer will understand, it will be necessary to know the 'command' words. These can vary from program to program, but for this one we can use an example which has been available for several years. Although it has been updated so that it can be used with most of the computers found in primary schools, its command language remains the same.

Once you have the program box on the screen, you can enter the appropriate commands. Remember, that for this program, there is a limit of twenty lines, and only one command is allowed each line. Using the appropriate commands, your program will look like this:

REPEAT FOREVER (This is self-explanatory!)
WAIT UNTIL INPUT 1 IS OFF (When the light sensor is covered)
SWITCH ON 1 (The light is connected to output 1)
WAIT 3 (It is on for 3 seconds)
SWITCH OFF 1 (Then it is turned off)
WAIT 2 (It is off for 2 seconds)
SWITCH ON 1 (Then turned on again and so on)
WAIT 3
SWITCH OFF 1
WAIT 5
SWITCH ON 2
MOTOR D ON (This is connected to two outputs to allow the polarity to be reversed)
POWER D 5 (This is half power – the speed control)
WAIT 3
SWITCH OFF ALL
WAIT 5
END REPEAT (Combines with the repeat command to show the commands that are to be repeated)
END PROCEDURE (Ends the complete sequence of commands).

Using control technology with children

We have already mentioned that control technology is one area of the curriculum that is often overlooked. This is not just the perception of the authors, but it is born out in many OFSTED reports. Whatever your feelings about these, they can at least highlight areas in the curriculum that still need to be covered. We do not believe that the only reason for this is the cost of materials. More often it is the old story of lack of confidence on the part of the teacher. This is quite understandable; science and technology both require a considerable amount of preparation, and, often due to a lack of communication on the part of scientists themselves, are still areas of mystery to large sections of the population. Nevertheless, it is vital that our children should be introduced to them at the earliest possible stage of their education.

We are aware that what has been described in this chapter is work more appropriate for Key Stage 2 children. However, if children are allowed to progress through a carefully thought out science curriculum with an appropriate use of information technology, then they will have already used and understood many of the procedures described above. Progression should never be the dry repetition of what has gone before, but a reinforcement of ideas "*through the increased complexity of learning tasks, the linking of the various procedures as a whole process and the gradual withdrawal of support, with a growing expectation of independence*".

It is quite possible to begin the work of control technology in Key Stage 1, and indeed one of our case studies shows how even nursery children can be introduced to the idea. One of the essentials inherent in (although not exclusive to) these control activities is for the child to make decisions based on previously collected data, and to predict what will result from these new decisions. This 'hierarchy of ideas and decisions' can occur from discoveries made in the sand tray just as much as those made in the sixth form laboratory. We have observed five-year-olds making decisions about the material properties of sand and water, which are just as sophisticated in their own way as those made by sixth form pupils when studying the concept of density.

Amongst the plethora of attainment targets, league tables and the like, the necessity of introducing children to new concepts at the appropriate developmental stage in their maturity often seems to be forgotten. There must be a natural progression of learning in all subjects, and as Philip Stephenson (1997) points out, control technology has a very structured hierarchy of its own. In his summary of this progression he lists the processes in relation to commands given to a model robot, such as 'Roamer'. These self-contained robots should be familiar to most primary school teachers. They are often used to enhance mathematics teaching along with computer screen 'turtles', but are also a natural precursor to control.

They rely on a series of basic commands such as FORWARD, RIGHT (for turn right) and BACK. There are no measurements involved, only arbitrary numbers. A series of commands could therefore include:

FORWARD 5
RIGHT 1

If the children were to program this in blindly and then press GO, they would have no idea how far FORWARD 5 would take the Roamer. Nor would they realise that a turn to the right of 1 would hardly be noticeable. It is therefore obvious that using Roamer itself requires a progression of learning procedures. The children could write their own simple programs, and use a partner as a robot. They will have learnt about quarter turns, or half turns as an introduction to angles or even as an introduction to the points of the compass.

It is not a great leap for them to accept that the command RIGHT 90 represents a quarter turn. As for what the command FORWARD 5 means in terms of distance, hopefully children can be persuaded to measure FORWARD 1 and go from there!

This progression of control sequences for the use of the Roamer can also be seen as the processes involved in all areas of control technology. The children may start with a simple isolated command to make the robot go forward, but quickly progress to repeat commands or programs which involve flashing lights and musical sounds.

Case study 1: use of a ladybird floor turtle in a nursery classroom

Many thanks to Kerrie Albon, and the children of Willow Class, at Westbury Early Years Unit, Letchworth, Hertfordshire, for showing us how even the youngest children can experience, and begin to understand, the basic concepts of control technology.

The class were learning to program the Ladybird. This robot is a simpler version of the Roamer which is described elsewhere in the chapter. The Ladybird has four simple arrow buttons to control direction – backwards, forwards, left and right. These buttons also dictate the distance to be travelled. One push for a given distance, two pushes for twice the distance, three pushes for three times, and so on. It is voice activated so that it will count aloud the times the arrow button is pushed. In the middle of the four arrows is the start button.

Figure 6.10 shows two children using the Ladybird. (Contrary to appearances, girls were heavily involved in this activity too!)

The children worked in groups, both boys and girls. As they gathered around the Ladybird together with their teacher, they were happy to have their voices recorded on a tape recorder. As any teacher knows who has used a tape recorder in their classroom, the results never seem to match what you know actually went on! Nevertheless, it was obvious that even at four years old, these children were beginning to realise that the Ladybird was something more than just a

Figure 6.10 Two children using the Ladybird.

plaything. Not that they didn't have fun! Quotes taken from the tape were often punctuated with sounds of laughter – quite right too!

Teacher: *How are you going to make it do more than one thing?* (i.e. after she made it go forwards).
Thomas: *Make it go backwards.*
Teacher: *What happens when you press the button more than once?*
Jacob: *I pressed it three times, and it goes more.*
Dillon: *I press it seven times.* (Sounds from the machine as it counts aloud how many times the arrow button is pressed)
Teacher: *What do you do to make it work?*
Ben: *Press the middle button. I'm going to press forwards.*
Teacher: *That's very good, well done.*

There are changes within the group as some leave and others join. For the sake of the newcomers it is explained that after pressing the arrow buttons, the central button needs to be pushed to make the machine move.

Dillon: *Press other arrows.*
Ashley: *Did it go a different way? Why is it going round first?*
Thomas: *Can we make it go backwards?*
Jacob: *Press arrow more than once.*
Ben: *It went that way.*
Keziah: *I'm going to press forwards.*

Tasmin: *You can share with me. Press for going backwards.*
Keziah: *It won't go backwards.*
Keziah: *I know. I'll show you what it does.*

At their first play with the Ladybird, one aspect that the children found puzzling was that it did not go backwards or to the side when the relevant arrows were pushed. If the 'backwards' arrow is pushed the Ladybird spins round on its axis and immediately travels forwards – in the direction that it is now facing. To a large extent this initial disappointment was overcome by simply suggesting to the children that the Ladybird (unlike the Roamer) was a model of an actual animal. Just like us, a ladybird would not choose to walk backwards, but would turn round to face the direction in which it wanted to move.

This raised another interesting teaching point. These were very young children, so how were they to decide and explain in which direction the Ladybird should move? Left and right would be difficult and at this age compass points and angles out of the question. However, this problem was soon overcome by using the structure of the classroom itself. Hence the directions would be: move towards the wall with the door, or towards the window (the opposite direction).

Figure 6.11 Drawing of a Ladybird produced by a member of Willow Class.

Figure 6.12 Another drawing of a ladybird produced by a member of Willow Class.

For left and right, then the wall with the sink, or the wall with the green board, were just as good.

This very enjoyable experience was not only recorded on tape, but also by pictures taken using the school's own digital camera. For good measure, the children also added their own pictorial record. Two examples of these are shown in Figures 6.11 and 6.12.

Case study 2: using *Robolab*

Our thanks to Class 5 at the Tewin Cowper Endowed Primary School, near Welwyn, Hertfordshire, and their teacher Paul Edwards for this case study.

Tewin is a small rural school, and Class 5 consists of thirty-one children from Years 5 and 6. They had already been introduced to the theory of control in a previous lesson, and had practised as a class the basic programs for the *Lego Mindstorms* and the RCX brick, using the *Robolab* software already installed in their computers. Several of the children were already conversant with these programs, and had little difficulty installing basic programs to control the robot. The school follows the QCA scheme for ICT, modified to fit the school's own requirements and needs. Thus control was being taught in Year 6, but children would have had some previous experience of programming during Years 2 and 4, using a 'floor turtle' and a 'screen turtle' respectively.

In this, the second lesson, the children took turns to visit the computer, in order to try out more advanced programs. They were organised into groups of three at a time, with a classroom assistant to help them. The computers are situated just outside the classroom, so there was easy access to the rest of the class, who were working with their teacher either completing worksheets associated with the project, or writing up the programs that they had used. At this time, because of the other calls upon the school's budget, there was only one robot available. Nevertheless, several groups were able to work at the computer during the available time, and it was estimated that only one more period would be needed to complete the process.

Each group of children soon progressed from the simple Pilot 1 program, through Pilot 2, to the more advanced Pilot 3. For the most part they did this instinctively and without difficulty, as they all wanted the robot to do more than just go backwards and forwards, or just spin round. For instance, the second group, by checking the meanings of the various icons, realised that they could easily fix the light to the robot. They searched through the box for the light, discovered how to fix it to the robot, and programmed it to flash on and off as

Figure 6.13 Cover designed by Amy to illustrate her robotics work.

> Programming the vehicle:
>
> *Tom, Lucy and I firstly programmed the RCX using pilot level one. We made the RCX turn around for eight seconds, we found out that the motors run at different speeds to make it turn, we found this fairly easy, we also did similar things with pilot levels 2, 3.*

Figure 6.14 Writing about the robot.

the robot moved along. Another group planned a route for the robot to follow, and also successfully completed a program for this.

A cover designed by Amy to illustrate her robotics work is shown in Figure 6.13, and Figures 6.14 to 6.16 show examples of writing about programming the robot.

> Robotics
>
> Programming the vehicle: On robolab we used Pilot level one to program the R.C.X. Pilot level one enables one port to be active and to determine the amount of time it is active. You can also program the power and direction of the motor. We also did similar on pilot levels 2, 3 and 4.

Figure 6.15 Writing about programming the robot.

> Programming the vehicle:
>
> Will, Tom and I, firstly programmed the vehicle using pilot level one. We made the RCX turn around for eight seconds. We found out that the motors run at different speeds to make it turn. We found this fairly easy. We also did similar things with pilot levels two and three.

Figure 6.16 Describing how the robot was programmed.

Two factors of particular interest emerged. It was necessary to ask the children to look carefully at the programs that they had written, and to think what they had actually asked the robot to do. In this way they had to mentally put into their own words the actual program, and check to see if it was possible, and to see if the commands reflected exactly what they wanted the robot to do. This was a very similar process, albeit carried out away from the computer, to the actual writing of commands for the *CoCo* program.

The second point of interest concerned the social interactions within each group. Because, computing of this kind is often seen to be 'a boy's thing', the boys in the groups tended at first to monopolise the computer. However, when a problem arose, it was more likely a girl who would find the solution, and thereafter take control, or at least take an equal part in the proceedings. This will probably not come as a surprise to older readers!

During a later visit, we were able to observe how some of the more ICT articulate children were able to cope with an advanced control project. For this they used one of the kits specially produced for this control project by Lego. On this occasion the children chose the 'Ghost Ride' kit. This involves a rachet drive to move a trolley, and a set of cog wheels for gearing purposes to slow down the rotation of the motor. Building these and understanding how they worked was a valuable science lesson in itself. The model also includes a light (the ghost) and a small plastic model of a skeleton which springs up when the trolley hits the buffer at the end of the rachet.

> The ghost train
>
> *Lucy, Bill and I programmed the ghost train on amusement park Ghost ride 2. We made the train up and down the track. At the same time we made the ghost light up and the skeleton pop up for two seconds. We only did one step out of eleven we found it hard at first, but we got used to it. We then printed out the front panel from the computer.*

Figure 6.17 Describing the Ghost Train.

> *Programming the ghost train: On robolab we clicked on pilot level 4 then on amusement park and finally on ghost ride d. This level had a preset program which you could download and run straight away. You could also add more steps to make it do more things.*

> The Ghost train
>
> *Will, Billy and I programmed the Ghost train on amusement park, Ghost ride 2. We made the train go up and down the track. At the same time we made the ghost light up and the skeleton pop up for two seconds. We only did one step out of eleven. We found it a bit hard at first, but then got used to it. We then printed out the front panel from the computer.*

Figure 6.18 Two children's desciptions of programming the Ghost Train.

Figures 6.17 and 6.18 show examples of descriptions of programming the Ghost Train.

There is a set program for this kit, which is simply activated by pressing the 'Ghost Train' icon. Whilst this might be of value for the less able pupil, this group soon decided to build their own set of commands, using one of the advanced and more versatile *Lego Mindstorms* programs. Without difficulty, they soon had the model working with sound and flashing lights included. There was very little else they could have added to this model, for it had been made to a specific design. However, there was, subsequently, much interesting discussion, for the children by this time were keen to design and build their own model with a variety of inputs. Even at this stage they realised that they would first need to discover if this system was flexible enough to allow them to do this.

Conclusion

At the start of this chapter we explained how control technology should evolve from children's previous experience of building and using mechanical devices. At the same time as children are using these, they should also be having the opportunity to enjoy the parallel experience of developing their own control language through the other activities described above. Cross-curricular experiences, especially the use of a floor or Roamer Turtle, as well as LOGO, can help to develop children's control technology experiences, and conversely, ideas and attitudes developed through control technology can be applied across a number of other curriculum areas, especially investigation techniques.

The purpose of this chapter has been to illustrate not only what control technology is, but also to explain why it merits a place in the primary curriculum. According to Stephenson (1997), it gives children an understanding of an expanding workplace or real-world application and it also offers context for the development of skills and attitudes, especially investigating and problem solving. These are both important areas. As we mentioned at the beginning of this chapter, there are very few things in society today that do not have at least a small element of control as an integral part. It therefore becomes important that children understand how computers and microchips provide this control. Whether it is a simple, direct instruction, perhaps something as simple as using an on/off switch, or a more complex programmed response, this understanding is essential in today's complex society. We have already mentioned in other chapters the importance of developing the key scientific investigational skills of hypothesis, prediction, observation, recording and the drawing of meaningful conclusions. Control technology plays a very important role in this as well as allowing the teacher to encourage the children to engage in open-ended dialogues, crucial to effective learning. As Stephenson goes on to say, the use of this involves greater intellectual involvement from the child which should correspond to their stage of cognitive development. As with all aspects of primary science, the teacher has a key role to play here. The teacher needs to be able to recognise the progression, skills and concepts that are part of control technology and to match learning experiences to the child appropriately. Once this has been done, then the children will have access to an area of the curriculum which is genuinely exciting, powerful and innovative.

Reference

Stephenson, P., Chapter 3 in Information Technology and Authentic Learning, edited by MacFarlane, A., Routledge, London, 1997.

Chapter 7

Museums, zoos, gardens, art galleries and ICT

At first glance the place of museums, zoos, gardens and art galleries might seem to be unlikely subjects for a book of this kind. However, because they can be closely connected with science education, they will also have close links with appropriate ICT. Although a school will have many different reasons to pay a visit to one of these places, the one that should concern us is a visit as part of a science project. Whilst zoos and gardens may not be so readily accessible, most schools will have access to at least a local museum and art gallery. Consequently, we will refer mainly to these in this chapter, although much of what is written could easily be applied to zoos and gardens as well.

Although a topic approach to science teaching is no longer thought to be appropriate to the needs of the National Curriculum, it is unlikely that a museum visit would be undertaken without some direct educational application, project or otherwise. Of course it is an enjoyable experience to explore these places, and have the chance to wonder at the variety of exhibits on view. However, this is an approach more appropriate to the individual, than to a class of thirty or so primary school children, and for these some planning and organisation is needed. This is an effort well worth making, for as is frequently repeated throughout this book, there should always be the opportunity for children to have offered them first-hand, practical experiences. Quite apart from the fact that all of these places are better resourced than any school could ever hope to be, there is little to compare with an organised, focused visit to any institution where 'real' scientific objects can be seen. The opportunities presented by such a visit are enormous. The possibilities to examine close-up famous and historically important artefacts such as Stephenson's Rocket, the Gloster Whittle E/28 (one of the world's first jet aircraft), dinosaur skeletons in the Natural History Museum or rare animals in a zoo or wildlife park, can provide memories and levels of understanding and appreciation that would otherwise be impossible. Often children are surprised at either how impressive or unimpressive the real artefact is when compared with an image. For example, the popular image of Stephenson's Rocket is of a large bright yellow steam engine, yet when viewed close-up today, the original is small and black with a very rough outer casing. In short, the reality bears little resemblance to the popular image. Sometimes this can only be truly appreciated with first-hand experience.

Links to the curriculum

Topic or no topic, it is likely that a visit to a science museum will be part of the ongoing curriculum. The teacher will have decided when the visit should take place. It may start off the project and therefore help motivate the children for the work to come. It will provide the children with first-hand experiences of the concepts and ideas that will follow, and will allow the children to make sense of these ideas that in school might otherwise appear abstract. It may complete the project, and therefore help the children to understand the concepts of the science that they have been studying. If the class is about to begin to learn about 'Forces' (National Curriculum Sc4, Key Stages 1 and 2) then a museum visit will enable them to see 'concrete' examples of how these otherwise abstract forces work and are applied. A visit at the start of such a project would enable the teacher to remind the children later that they had actually seen these abstract ideas at work. Conversely, if they have been learning about plants and animals (National Curriculum Sc2, Key Stages 1 and 2), then a visit at the end of their work would not only enable them to see some examples, but more importantly, help place them in their correct ecological framework.

Organising and managing the visit

The teacher will have considered how best to make sure that the children know what to look for during their visit. In most museums there is so much to look at that unless some guidance is given, even adults find it difficult to take in this plethora of information. The guidance may take the form of a questionnaire or worksheet. Many museums provide these for a small price, but teachers, perhaps with the help of the class, may prefer to design these themselves. This will give an obvious opportunity to incorporate some good ICT from the very start, and during the planning stage itself. The children may use a word processing or a desktop publishing package to produce these. This is a good opportunity to develop literacy skills such as the use of appropriate and relevant questioning, as well as to give the writer a sense of audience. Obviously, graphic design considerations of layout, colour and images will be very important. Through the use of clip-art and digital imagery the children can create illustrated worksheets, which will later help them to find particular artefacts in the museum. Of course, with the guidance of the teacher, the children can produce worksheets that are specifically geared to that particular class or group. Although material that is commercially produced by museums is normally very good, often designed by education specialists, this tailoring to individual groups or classes is necessarily absent. The children now take on a new role; they may still have to complete a worksheet, the traditional burden of any child on an educational visit, but as they have to produce something that they themselves will use, there is added interest and motivation.

A sample worksheet is illustrated in Figure 7.1, including clip-art that has been downloaded from Microsoft's online clip-art gallery, which can be visited at http://dgl.microsoft.com/?CAG=1. The clip-art for the Spitfire and Concorde have been turned into a watermark by using the picture toolbar, clicking on the 'Wrap text' button and then putting the image behind the text, then clicking on the 'Image control' button and selecting 'Watermark'. This reduces the colour so that the image does not block out the text.

This is an example of a possible worksheet for a topic on 'Flight', and presupposes that the class has prepared themselves for the visit; hence the mention of the class time-line. It could be based on a visit to any suitable museum. It has a mixture of specific, closed and open-ended questions, which can be designed to cover many areas of the curriculum. 'Flight' used to be a favourite primary science topic. It is a shame that more time cannot be allowed for it now, as children not only find the subject interesting, but also their enthusiasm with the practical work involved enables them to understand quite advanced concepts. As Figures 7.2 and 7.3 show, certain of these abstract concepts – in this instance the applied concept of 'Forces', are still in the National Curriculum. However, without 'Flight' an opportunity to explain them may be missed. It would not need complicated practical work. A simple paper dart would do, especially as the importance of flaps and ailerons could be explained at the same time.

Using the Internet to support educational visits

When planning and preparing any educational visit it is important that the class teacher visits the venue themselves first to get a 'feel' for the place and what it has to offer. No successful visit can take place without this. However, these days there is an extra source of information. The focus of the visit, and thus the content and focus of the worksheets, can be informed by visiting the museum's web site. These are usually excellent and should be the first place to go in order

Flight

1. During our work at school we looked at the history of flight, and particularly how people used, balloons, gliders and even kites to make flights. All these happened before the invention of an engine to power the aeroplane. Look for some of these amongst the exhibits, and write them into the table below. The first one is done for you.

Name	Date	Place	Balloon	Glider	Kite	Other
De Rozier & *D'Arlandes*	*1783*	*Paris*	*Montgolfier Hot-air*			

2. We shall enter all your work into the class time-line when we return to school. Look amongst the exhibits to find the first aeroplane to fly with a petrol engine and a propeller.

   In which year did it first fly? ....

   Who was/were the inventor/s and pilot? ....

   Is the exhibit a model, a full-sized replica or the real thing? ....

3. Look amongst the exhibits for the first jet powered aeroplanes. (There may be more than one type). What are they called and in what years did they first fly?

Name of Aeroplane	First Flew	Country

4. Look amongst the exhibits, and find:

   The first aeroplane to fly across the English Channel ....

   The first aeroplane to fly non-stop across the Atlantic Ocean from the USA to the British Isles ....

   The first passenger aeroplane to fly faster than the speed of sound ....

   Can you add one more "first" to this list? (Don't forget to put the dates down for our time-line).

   Which of all the exhibits that you have seen today would you most like to fly in? Tell us why, and draw a picture of it on a separate piece of paper if it will help.

Figure 7.1 Example of a worksheet for a gallery in a museum on the theme of 'Flight'.

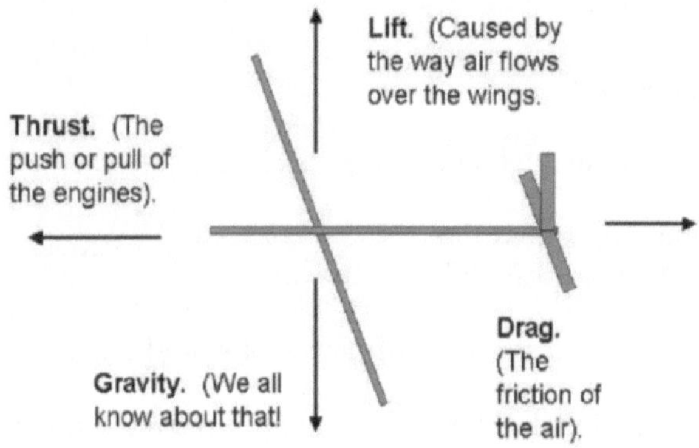

Figure 7.2 Diagram to show the balance of forces that keep an aeroplane in flight.

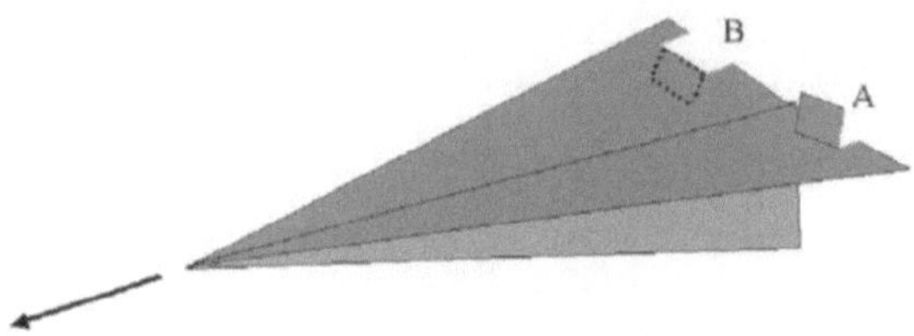

Try a paper dart with one aileron up (A) and one aileron down (B), to see how it flies

Figure 7.3 Diagram to show how the ailerons could be positioned.

to find out about the museum and what it has to offer. The larger, internationally renowned museums usually have superb sites with an excellent range of online information and resources. A good web site should be easy to navigate, in that it will be easy to find information, the links will be obvious and the screen layout will be clear. It may also have multimedia features, in that it provides sound and moving images to go alongside the text, the static images and the graphics. A list of recommended web sites is provided at the end of this chapter.

Most of these sites have several main areas, and usually include a section for educational visits, giving contact details and services that are on offer. This is a good starting point for getting the basic information such as opening times or charges (although relatively few of these places currently charge for admission). They will also have details about current exhibitions, both permanent and temporary, and will often have virtual tours, where the galleries have been captured and digitised into a sequence of navigable images. By pointing, clicking and dragging, the user can take a tour of parts of the museum or gallery via their computer. This of course should never replace a real visit, but it enables the teacher and children to plan exactly what they want to see, utilising the time spent in the galleries effectively. A web site also provides excellent support when following up a visit, in that it gives the children an accurate reminder of what they have seen and where they have seen it. If this is a museum or gallery in another part of the world the virtual gallery tour provides the children with an opportunity that would otherwise be unavailable. At least the child can get a flavour of what this exotic distant location can offer.

In this context there are distinct advantages in using the features that the Internet provides. The use of the World Wide Web enables children to directly access resources from anywhere in the world that are online. Although this is not the same as actually visiting for real, it does, however, allow the user to bring into the classroom materials and resources that would otherwise be unavailable. From as far afield as museums and galleries in North America, Europe or Australasia, children can directly view, compare, and contrast artefacts from each. Although this has always been possible through the use of educational broadcasts, especially when a video is used, there is an immediacy that the Internet provides. If children are working on a particular project or topic, they can go online and by using a search engine, can have access to any relevant information quickly and easily. This can then be integrated into a range of other applications, but it should be remembered that any text or images are of course subject to the laws of international copyright.

Primary science and the National Grid for learning

A further feature concerning educational web sites that is offered to schools in the United Kingdom is the National Grid for Learning, usually referred to as the NGfL. This was launched in 1998 to provide a virtual learning community by allowing schools to directly and easily connect with a wide range of learning institutions such as museums, galleries, libraries, colleges and universities through the provision of the NGfL web site. It was developed to co-ordinate the vast amount of high-quality British material on the web. This acts as a portal site, which enables access to a wide range of web sites and to communicate via the Internet and email. Although these sites can be accessed through typing in the web site's address directly if it is known, the use of the NGfL site allows the user to select the required museum or art gallery by simply pointing and clicking on the appropriate link. One of the main advantages of the NGfL is that all participating sites carry an NGfL logo, which signals to the user that the site has been 'vetted' and that it is suitable for use in school as it contains high-quality educational resources. This same facility also enables all kinds of specialists to be available to schools via email, with a discussion board or chat room or a 'submit question' option on the web site. So, for example, a pupil with a scientific query could email an expert at a museum anywhere on the grid, knowing that it should be in complete safety as it is an official NGfL site.

This has many advantages for primary science. Teachers can share and discuss ideas from their schools or homes via email or discussion boards. A wide range of relevant learning materials are always available online which can be accessed from home, a library or any other location where there is a computer that is connected to the Internet. Because of the sheer volume of material available online, it is always possible to access material to suit children of all ages, abilities, or needs, and also allows for access to opportunities for wider life-long learning. Many of these resources have been written and used by teachers with their classes, so they have the advantage of having first been tried and tested with real classes in schools. It is this ability to connect teachers and learners together in such an easy and immediate way that provides a powerful opportunity for all. It also ensures that children and their teachers in large urban schools and those in small, isolated rural schools will have equal access to these resources – as well as each other. Being a British National Grid for Learning, it ensures that these materials are specifically geared towards the National Curriculum. As a natural extension of this, many British Local Education Authorities (LEAs) have also set up their own regional and local grids, where material, advice and easy contact with advisory staff are made available for local schools with a local emphasis.

The Virtual Teacher Centre (VTC)

A corresponding development is the Virtual Teacher Centre. This is another web site portal that is designed to allow teachers, subject specialists and co-ordinators, senior managers, Headteachers and LEA Advisors to share information and advice about specific subjects or aspects of education. It uses bulletin boards, and there are curriculum sites that contain schemes of work, medium- and short-term plans, lesson plans, worksheets or even interactive, on-screen activities. There are also links from here to a range of government education web sites such as the Department for Education and Science, the National Curriculum, the Qualification Curriculum Authority and the Schemes of Work which are located on the standards site. Another excellent site is *teacherxpress.com* which performs a similar role to that of both the National Grid for Learning and the Virtual Teacher Centre. Provided by a leading educational software supplier, it contains all of the features described above as well as a great deal more, but of specific interest to the primary science teacher is the section of resources for science at Key Stages 1 and 2.

ICT within educational establishments

However, although this virtual community provides an excellent resource to the primary teacher, it is also useful to see what ICT the museum itself has to offer. Whilst much of this may not be suitable for primary schools, many museums offer access to a variety of databases. These may be to help identify animals or plants, and show how and where they live, their place in the environment and the food chain. They also help the user to find examples amongst the exhibits, and may even involve the children with opportunities for some practical 'hands on' science. A particular favourite of the authors' is a visually displayed branching database which enables the user to identify various examples of rock, using a series of questions with yes or no answers. Instead of using a simple display screen, there is a pathway in the form of branching tunnels. A ball-bearing rolls down the tunnels. At each branch, depending on a 'yes' or 'no', the ball travels right or left until it ends up in the appropriate box. Although this may look like a simple game, it is computer driven, based on the binary system.

We have already seen that a visit to an art gallery may suggest uses for ICT that at first sight have little connection with science. However, we have often in this book looked for opportunities to link different areas of the curriculum, and art and science should be no exception. There are some obvious paintings that can be used as a starting point for a science investigation. The paintings by Joseph Wright (of Derby) which show scientific experiments in progress are an obvious example. However, there are many less obvious examples such as those showing early steam locomotives or watermills. Even paintings such as Bruegel's 'The Tower of Babel' provide opportunities to explore early building methods. Some of the larger art galleries also provide computer suites where the collections may be accessed online from their own network servers, so that the user can find out further information about a particular painter or painting, or where exactly in the gallery these are located. Often these are available online via the web site, or are sold on CD-ROMs so that they can be taken away and used in school or at home. These have the advantage of ensuring that the children can continue any scientific investigations in school, but again within the laws of copyright.

Hands-on science

Many museums have galleries that allow children to have 'hands on' experience of the exhibits. These are often working models that can incorporate such things

Figure 7.4 Type 1 lever, such as is found on a see-saw, or when lifting a heavy object.

Figure 7.5 Type 2 lever, such as that found on a wheelbarrow.

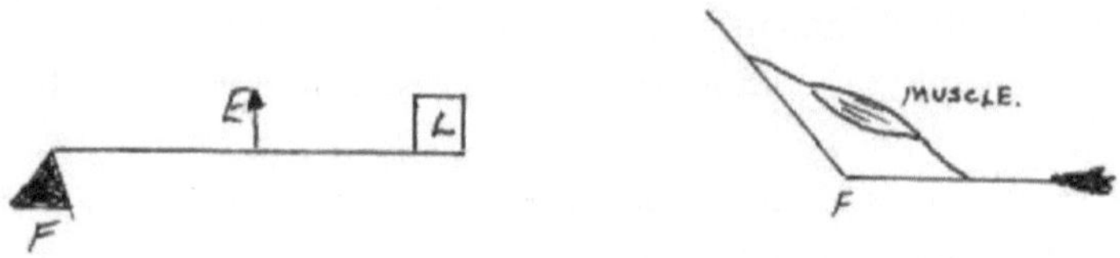

Figure 7.6 Type 3 lever, such as that found on the forearm.

as levers, pulleys, or telescopes and mirrors. Some even have various structures, for example an arch bridge for children to walk across. All these can be great fun, and it is right that children should experience them. However, after the children have used them, do they remember what these models did, or how they worked? The children may have only played on them in just they same way as they would on the swings in the local park. Perhaps this may not matter, but sometimes if there is no follow-up to the activity, then an opportunity for real learning may be lost. This would be a great pity as so many of these models incorporate areas of mechanics and physics that are often difficult to replicate in the classroom.

Levers

Nevertheless, once they are back in their classroom, it is often possible for the children to construct some of these models. Levers are relatively easy to make and understand. There are three kinds, depending on the position of the various parts, and these are detailed in Figures 7.4 to 7.6.

Whilst the exact positions of the fulcrum, load and effort can change, there can only be three variations. Probably most levers in use are of the first type. The load and effort on the see-saw changes with every movement. The axle on the wheelbarrow is the fulcrum, and it is the muscle that provides the effort on the forearm. These can be made as replicas of this diagrammatic form in wood or card and displayed on card, or the children can make the actual models themselves.

Pulleys

Pulley systems are difficult to build in the classroom. The kinds of pulleys that are found in various educational catalogues are small and fiddly. However, if children were able to use a large block and tackle in the museum, then they can learn more later by reference to those used in workshops, the building trade, or to lift the sails on yachts. Many practical activities can be found in a variety of sources to follow up those that children have experienced in the museum. Experiments with different lenses will help children to understand how the telescope works, and mirrors can be used to explain the properties of light, as well as to make models of periscopes and kaleidoscopes. Even the Archimedean screw, examples of which are often found amongst these exhibits, can be replicated in the classroom.

How to make an Archimedes' screw

You will need: a length of transparent plastic tube, some heavy-duty plastic sticky tape, and a plastic or metal cylinder (a plastic bottle will do). Use a fairly wide-bore plastic tube. The curved arrows in Figure 7.7 show the direction of rotation for this model.

Wind the plastic tube round the cylinder (bottle) and fasten at both ends with the plastic tape. Then place one end of the cylinder into a sink of water so that the open end of the plastic tube is under the surface and rotate the cylinder complete with plastic tube. If a little red food dye is mixed with the water it will make it easier to see the water rise up the plastic tube. Try different-sized cylinders with differently spaced tubes, such as the one shown in Figure 7.8. Do they all work in the same way?

Note: Any screw, whether it is an Archimedean or a simple wood screw, is a kind of helical version of the inclined plane. The wedge-shaped inclined plane, or slope, is one of the three basic kinds of machines. The others are the lever and pulley. When you turn a screw, you are for all intents and purposes, pushing a weight up a hill. (National Curriculum Sc4 Forces and Motion, Key Stages 1 and 2)

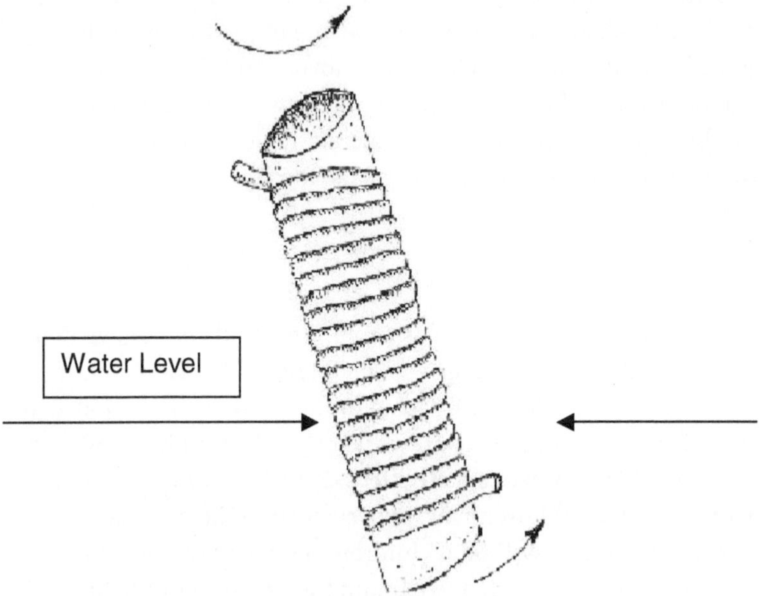

Figure 7.7 Archimedes' screw. Use a fairly wide bore plastic tube. The curved arrows show the direction of rotation for this model.

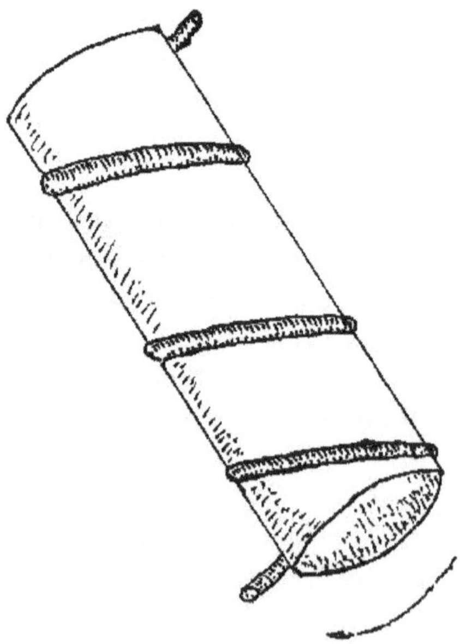

Figure 7.8 Try different-sized cylinders with differently spaced tubes.

All these machines are often found in the 'hands on' sections of museums. A combination of them is included as one exhibit in the Science Museum in London. As well as two Archimedes' screws, it has pulleys, lifts and a moving belt. The whole system is designed to move loose grain round in a continuous cycle. Children can feed the grain into the cycle and work the various machines. As part of a follow-up exercise the whole system could be drawn on a computer screen using appropriate drawing software, or even just be using the drawing options within another application such as *Microsoft PowerPoint*. This will not only enable the teacher to explain the way the various machines work, but the system can be animated in a way that an ordinary drawing cannot.

Using educational multimedia

We have frequently stated throughout this book that there is little to compare with the ability to have practical, hands-on activities such as those described immediately above. However, through the use of good interactive web sites and CD-ROMs, the children can not only have these worthwhile and meaningful experiences in the classroom, they can also build upon these experiences by the use of multimedia, perhaps via the museum web site itself. Multimedia allows text, static graphics and images to be combined with sound such as speech or music and video clips of moving images. This is a very powerful option indeed, and as such provides a huge amount of scope for the primary scientist. It enables children to read, watch and listen to video clips that are produced on CD-ROMs from a wide range of publishing companies, as well as the museums and galleries themselves. These explain a whole range of scientific concepts and ideas to the children, and many of these are extremely sophisticated. They look smart, work quickly, have excellent graphics, the images move smoothly, have good sound quality and are often presented by familiar faces from the media. Whilst many of these are genuinely excellent, and provide a rich learning experience, it is important not to be taken in by slick presentation alone. As with all educational resources, the teacher needs to be aware of pertinent issues, many of which were raised in the previous chapter, and include aspects such as having a manual that is easy to read, understand and (very important for this type of book) have a good index. A CD-ROM for example, that graphically shows the blood circulating through a body in an animated presentation can bring a great deal to children's scientific understanding. However, if it is purely a static representation, or simply an on-screen version of a paper book then there is no value-added component and an excellent opportunity is lost. The power of a computer lies in its ability to handle a large amount of different types of information quickly. The ability to animate and to make images move is a powerful one that the best multimedia CD-ROMs utilise to the full.

Evaluating multimedia software for science – some points to consider

The teacher is best advised to ensure that the children have prepared worksheets, or at least a set of notes, to enable them to navigate through web sites and CD-ROMs. These will help them to find the essential facts and features connected with the ongoing topic. Without this guidance, children so often waste both time and opportunity in searching through irrelevant information. For example, if it is the circulatory system that is being studied, then a worksheet should require answers to a set of relevant questions in order to keep the children focused on the task in hand. These worksheets may need to include some instructions on where to find the information within the CD-ROM or web sites themselves. If children are asked questions about Greek science, they should not be distracted by Roman engineering (or at least not until the questions are answered!). Although all web sites are different, some basic principles need to be followed. They should identify where to point and click, perhaps giving some key words to assist navigation. They might be presented in the form of a flow chart, where there are arrows to/from key words, phrases or buttons.

It is also important to bear in mind other key considerations when selecting multimedia CD-ROMs or web sites. The teacher needs to be aware of any bias that may exist. A developer may have certain values and attitudes that are consistent with a particular commercial or national viewpoint. This is a particular danger if a piece of software or web site is aimed at a specific market. If it is aimed at a large domestic market in one part of the world, it may not directly appeal or be relevant to a school somewhere else. There may also be difficulties with the language. It may not be appropriate for the age of the children. Some museums may not have English language versions of their web sites, whilst others that do may have poor translations. There can be other variations, such as the use of American English on web sites or CD-ROMs. These will have significant word or spelling differences, and this will mean that the teacher needs to be aware of differences and alert the children accordingly. The Internet is a dynamic, rich environment and language differences should not prevent users in one part of the world using material from another, especially as some of this material can be highly original and very useful. However, it is also important to be aware of the fact that material that is freely available and legal in one country may not be so in another. The school needs to ensure that any material that is accessed is both safe and legal in this country.

Conclusions

Whether you are using CD-ROMs to enhance the school visit, or just to enhance the science being taught in the classroom, then they will inevitably include some science that seems to be outside the confines of the curriculum. A visit to a museum or gallery will certainly offer a wider view of science. As Fowler (2002) states: "There is a notion that 'If its not in the National Curriculum, why should we teach it, if pupils don't have to learn it?'" We can understand this point of view. Testing, SATs, League Tables, the general bureaucracy, all of these do not encourage teachers to reach beyond the National Curriculum. This is a great shame, for museums and galleries do not, hopefully, restrict their exhibits to a curriculum. Energy, for example, does not seem to appear even as a word in the National Curriculum Key Stages 1 and 2, yet does so in many museum exhibits. Perhaps it was thought that Energy was too difficult a concept for primary age children, despite many schools having successfully taught it for some time. Certainly all these abstract concepts are difficult to grasp if taught badly. However, is 'Energy' more difficult than 'Forces'? A visit to a science museum will allow the teacher to place such ideas as 'Energy' in their right context. Children will be shown examples of green plants utilising it to make food, almost it seems out of thin air. They will have the opportunity to see electricity, light and sound: all various forms of energy at work. Although 'Energy' may not have its place in the National Curriculum, ironically all these other examples do. After all, without 'Energy', none of these things, none of us, and not even the National Curriculum would exist!

Some relevant web sites

Although correct at the time of writing, please be aware that web adresses can, and do, change from time to time. The mention of a specific web address here does not guarantee its continued existence or location at this address.

www.teacherxpress.com – Portal site that links to hundreds of educational websites, including dedicated science sites and museums.

www.ngfl.gov.uk – The National Grid for Learning.

www.vtc.ngfl.gov.uk – The Virtual Teacher Centre.

www.nmsi.ac.uk – National Museum of Science and Industry, including links to the Science Museum, National Railway Museum and the National Museum of Photography, Film and Television.

www.ltmuseum.org.uk – The excellent web site for the London Transport Museum.

www.24hourmuseum.org.uk – Links to many other UK museums.

www.nhm.ac.uk – The excellent web site of the Natural History Museum.

www.rog.nmm.ac.uk – The Royal Observatory at Greenwich.

Bibliography

BECTa, *From Chalk Board to Internet*, BECTa, 1999.

DfEE, *Connecting the Learning Society*, HMSO, 1997.

DfEE, *Preparing for the Information Age – Synoptic Report of the Education Department's Superhighways Initiative*, HMSO, 1997.

Field, J.V. and James, F.A.J.L., *Science in Art: Works in the National Gallery that illustrate History of Science and Technology*, Monograph No. 11, British Society for the History of Science, 1997.

Fowler, P., 'Why Teach the History of Science in Science? (Part 1)', *Education Forum* (Newsletter of the Education Section of the British Society for the History of Science), June, 2002.

Chapter 8

Science, ICT and drama

At first glance there may seem very little connection between science, ICT and drama. However, as we said in our introduction we are writing as much about science as about ICT, and the ways in which both can be taught. Certain aspects of science lend themselves to drama, particularly if it is to be taught to primary school children. If done well, the children will not only understand the science involved but will be better able to comprehend the way science and scientists work. Furthermore, children enjoy this empathic approach, and it does seem a pity that enjoyment is not a more significant part of today's curriculum.

Some possible areas of study

One area of science that lends itself to a dramatic approach is the body's immune system. How, within the National Curriculum for England's (1999) section on Life Processes and Living Things, can this system best be taught in a practical way? Not easily, which is probably why it seems that it is not required in any detail. However, it is very important. If not taught in science, then it should be a part of the general health education of children. To explain the immune system by using drama is particularly interesting for young children. It takes little imagination to think of bacteria as an invading army. The defenders are of course the white cells within the blood system. If this is presented as a play, then children can be dressed up in suitable costumes. The bacteria can be dressed as monsters, whilst the white cells depending on their role would be soldiers, doctors and nurses. Vaccines produce antibodies, and these too can be portrayed in an appropriate way. There would be little need for dialogue, but there should be a narrator to explain what was going on.

Another area of science which seems to have been be neglected is its history. When children study magnetism, they will often be shown the magnetic fields by observing the patterns formed by iron filings. Will they at the same time learn that it was Michael Faraday who first studied these? Similarly, when studying light and colour, and by looking at a rainbow, will they be told about Newton and his experiments with prisms? We have often observed classes carrying out experiments for a topic about forces. Although they may well be discussing gravity and its affects, no mention is made of Galileo's experiments at Pisa.

Studying the lives of these famous scientists can not only show children how the subject develops, but will also give it a more human dimension. If they discover that, for example, Faraday had no formal education but became interested in the subject through the inspiration of Davey's evening lectures, they will also learn something of the romance and wonder of science. These quite proper emotions seem to have been casualties of recent approaches to learning. When children discover that Newton liked to play truant from school to carry on scientific experiments on his own, or how Galileo was persecuted by the Church because his science questioned the established order, they become interested in these characters. This interest can lead children to ask what they did and why they were famous. The science is then shown in a more human and relevant context.

Making the science meaningful

It is very easy to portray these historical characters in short plays suitable for a class assembly. These could either be devoted entirely to one character or to several at one time. For instance we can utilise the (at present) fiction of time travel. One of the class can play the part of a time traveller, perhaps taken from a TV character; someone from 'Star Trek' perhaps, or for an older generation, 'Dr Who'. This character would take us through a guided tour of time to meet various famous scientists. Each would explain why they were famous, and do one simple experiment to illustrate the point. Hence 'Faraday' could talk about electricity, 'Newton' could explain about the colours of the spectrum, and 'Galileo' could drop his weights from the Tower of Pisa! (Williams, 2000). Nor should the women in science be overlooked. Marie Curie is an obvious candidate, but there are many more as any study of the history of science will show. Scientists and engineers from Ancient Greece, the Romans and China, should also be studied, as should the Byzantine and other Middle Eastern cultures.

Obviously one of the difficulties with this approach is the time available. We are not advocating that the complete curriculum be taken up in this way. However, if each class in a school were to produce only one such play a year this over time would allow the whole school to share in these discoveries. If the class performing the play was also able to do the relevant science in their science lessons, then they could answer some of the requirements of the National Curriculum in Science, as well as some of the History and perhaps Geography ones as well.

Accommodating the history of science within the curriculum

What might be perceived in the current educational climate as one of the main difficulties with drama, is that it cannot be easily assessed and recorded in quantifiable terms. It cannot be tested, levelled and/or count towards league tables. It does not have any kind of written outcome and as a result, cannot be recorded. It could be argued that it occupies valuable time in an already overcrowded timetable, and many teachers feel that they cannot justify the time in teaching it – always assuming that they feel confident or comfortable teaching it in the first place. The fact that it represents an essential aspect of primary education in that it develops personality, confidence and character seems often to be overlooked. As does the fact that it can often be an important means of teaching primary age children knowledge and understanding in a way that enables them to empathise with the material being studied. Unfortunately these factors do not seem to rest well with a curriculum that currently seems to reject creativity in any form in deference to an obsessively narrow perception of literacy and numeracy. The use of ICT can not only overcome these difficulties but can extend and enhance the role and position of drama in the curriculum.

Why use digital video?

Another potential difficulty, as with dance or movement work in P.E., is that once an action or a sequence of actions has occurred it can never be repeated. Despite practice and repetition, which can lead to similar outcomes, whatever movement has been acted out can never be exactly the same. This is where the use of ICT can make a uniquely significant contribution. Through the use of a digital video camera and simple movie editing software such as *Cinematic* from MGI or *iMovie* on Apple's iMac computer, children can record their own work through visual means using equipment purchased on a relatively modest budget. They can then edit and re-sequence their films using the software in much the same way that they can cut, copy, paste, delete and re-sequence text in a word processing package. However, there is an extra dimension as they can also edit the sound-track by dubbing, altering or removing the sound, or even adding extras such as sound effects and background music. Not only is each unique action or sequence of actions captured, but they can also be modified and enhanced according to the children's own requirements. The children have complete ownership; it is finished when they say that it is finished. It is another way of getting their message across, and can be a new and incredibly exciting addition to the primary science curriculum!

What equipment do I need?

There are however technical considerations. When selecting a particular computer for this purpose, it is necessary to choose one with a large hard drive and a fast processing speed. Most modern computers should be up to this task, but when purchasing new equipment it is well worth bearing this in mind if there is an intention to use video-editing software. Video files are very large, and consequently take up a considerable amount of disk space. The children will get frustrated if their creative efforts are hampered or restricted by a computer that will not process their requested functions quickly enough or will simply 'hang' because there is insufficient free memory space. It is always useful, if possible, once the video has been completed, to save it to an external medium such as a CD-ROM as this avoids the hard disk becoming filled up with a number of these videos, as this in turn can affect the overall performance of the computer in question.

As a computer can only handle video in digital form, it is preferable to have a digital video camera, where the video is recorded to an internal memory card rather than a tape. Any video that is recorded on a camera of this type can be connected to a computer via a USB port. The video can then be downloaded straight into any computer that has the editing package installed, and is ready for immediate editing or even playback. This is quick and easy, and does not require the computer to be restarted. Although there is a slight difference in quality between this and the more 'traditional' analogue video camera, the digital camera is more flexible as the image is not 'finished' once it has been taken. Although analogue video cameras can be used, this involves a slightly more complex series of operations and may require the installation of a video capture card inside the computer. This will add to the cost, requires fitting and the camera can only be used on the computer which has the card fitted. There are USB versions of this hardware available, where an external card can be plugged into a spare USB port on the dedicated computer, but these can be expensive. Moreover, the conversion of the analogue sequence into a digital sequence means that there is an extra technical stage to go through, which makes it slightly more complex for the children to have to deal with. This detracts from the main focus of the activity which is the production of the finished video, and of course, the science element, the content of whatever has been videoed.

Organising and managing digital video in the curriculum

So how might this powerful, yet relatively cheap and easy to use application be used in a primary school context, and how might other aspects of ICT be incorporated appropriately into the process at different stages? As part of a topic on forces, a Year 5 class is going to put together a short piece of drama about Isaac Newton and his discovery of the force of gravity. This will then be presented as part of a class assembly, and a permanent record is required for use in the future. Because this record may possibly be used in a number of different ways, some kind of 'master' video will be required. The class is organised into groups of two to three children who begin by collaboratively researching different aspects of Newton's life, and drawing and collating information from a variety of sources, including science CD-ROMs and the Internet. By using a word processing package and incorporating the copy and paste process, they produce a piece of original factual writing. They might enhance the text with the use of appropriate clip-art, copyright-free or cleared images from a CD-ROM or the Internet. They might even produce their own images through the use of a digital still camera, or use a scanner to digitise photographs that they have captured using traditional film and paper methods. These could be experiments that they have carried out themselves, or images captured on a visit to a museum. This could then be saved as a finished piece of work in its own right for the children's books or files, for a wall display or perhaps even the school's web site, or even all three. The provisional and dynamic nature of the use of ICT means that the work produced by a computer is never entirely finished – it is only finished when the owner or owners of the work say that it is. It can be resaved in different forms, so there are three different versions of the same text, or it can be progressively saved so that content and form evolve through a series of distinct and apposite phases.

The children can then carry out a procedure that Monteith (1998) calls 'transforming text' from a piece of continuous prose to a play script by reformatting the text on the screen through the use of the cut, copy and paste functions. This of course does not necessarily mean that the original text version has to be deleted; it only needs to be resaved under a new name. A narrator has to be added, the 'TAB' key has to be used to format the text into paragraphs to enable the name of the speaker to be inserted into the left-hand margin and stage directions have to be inserted into the script in the appropriate places. Already many different skills have been employed for both scientific enquiry and the use of ICT, as well as developing higher-order reading and writing skills to address the requirements of literacy. The children have so far had to search and synthesise factual information from a range of sources, both printed and electronic, read for meaning, use a word processing package to collate all of the information discovered, insert images from a range of sources, then reformat the text, taking into account changes in both audience and purpose. And rehearsals haven't even started yet!

Once the drama sequences have been developed and refined, the next stage is to video the performance or performances as required. If they are fortunate enough to have access to more than one camera then different camera angles and techniques can be used, such as close-ups or cutaways. These can then be edited into one piece of continuous, real-time video at the editing stage. However, even in the likely instance of only having access to one camera, interesting and creative results can be achieved. If there is more than one performance of the piece, then each one can be filmed from a different angle and can then be edited into one film, to give the impression that more than one camera was used. Of course the children would have to ensure that they were dressed identically and the lighting conditions would have to be similar, but this does provide an imaginative solution to what otherwise might be a problem. This in turn could lead to a whole new debate about the processes that are important in any kind of television, film or video production, such as continuity, and as Millwood (2000) states, it is opening up a whole new kind of literacy.

In order for the children to be able to carry out their own editing, and for the purpose of this description – essential for this to be a 'complete' activity – it is assumed that a digital video camera is being used. Without the cost of the computer, which the school will probably have anyway, the children will be using resources that cost at most, a few hundred pounds, and that will be simple enough for them to use. As has already been stated the process will only be finished when the children themselves feel that it is finished. Previously this kind of production and editorial freedom would have been impossible, as it would have required a professional editor, possibly located many miles away to edit the film and sound. It would have needed access to an editing suite, costing many thousands of pounds, but the editor would not have the immediacy or the sense of ownership that the children have. They will have their own say in the final outcome or possibly outcomes. The video may be 'stitched' together in a number of different ways, which can be saved as separate files or pieces of work. It might have several alternative endings, it might be a 'finished' video in its own right, a video clip on a web page, or perhaps could be part of a *PowerPoint* presentation. Extracts can be taken, used and reworked to be separate sequences or videos in their own right. 'Stills' can be captured from the video to enhance written text such as descriptions, reviews or posters. The children can add sound effects or music, or dub on extra speech easily and quickly. These, like the still images mentioned above, can come from a range of sources such as a download as a WAV or MP3 file from the World Wide Web (copyright restrictions permitting), a CD-ROM, or the children's own sound or music recordings. Sequences can have transitions, fade-ins and fade-outs, and can be titled. The finished product can look quite professional, yet can be achieved relatively quickly and easily. This is done by dragging and dropping the images and other effects from a library, which are a series of windows containing different sequences from the captured video, into the storyboard. It is here where the editor sequences the video into the order that they want, then makes any necessary adjustments to overall colour. This final stage in the production process is known as rendering (Medley, 2002).

Teaching and learning considerations

So Science, ICT and Drama can quite happily coexist. Admittedly this is still very much an emerging technology, but as we have shown, it has enormous and exciting potential. Science delivered through drama remains an investigative activity with the data becoming the action and the script. However, instead of using a database or a spreadsheet to process the raw data into tangible information from which meaning can be drawn, video editing software is used. Nevertheless, the scientific processes remain the same. The children will need to be able to hypothesise in order to answer a research question and predict what they think it is that they will discover. They will need to be able to use searching and sorting skills in order to research the material that they are intending to perform. They may even use search criteria from other ICT sources such as the World Wide Web or a CD-ROM. This material still has to be sifted, collated and synthesised into a form that can be dramatised, which will mean taking decisions that are based on a knowledge of the relevant and appropriate scientific facts. They will then need to make links from their research through their drama to draw meaningful conclusions and thus be in a position to test their original hypothesis.

They need to use the video camera to record their drama. When this has been done the recording will then need to be 'finished' in much the same way as the first draft of a word processing document or the raw data collected from an investigation or the information collected from a survey. In the same way that a database is used to interrogate data and to draw meaningful conclusions from raw data, then the first take of the video becomes the raw data. The children will then need to be able to make similar decisions as far as the ICT element of the process is concerned. How will they present their final piece of video? This in turn leads to other key ICT, but still scientifically relevant questions. How will they edit it? How will they piece it together? What sequences will they use? Will they put it together in the same order? Do they need to go back and film some more sequences because some of the original angles are wrong? What music will be used? Where will it go – at the beginning? Playing as background music? What sound effects might be needed? Where will all of these come from – CD-ROMs, audio CDs, or the World Wide Web? Will it be self-written and recorded? What transitions will be used? What titling will be needed? The permutations are endless, but for all that they can be almost as important as the original objectives of the science activity.

Conclusions

It can be seen from the above activity that ICT can take many different forms. It does not necessarily have to directly incorporate a computer, or even be a primary part of the activity being undertaken. In this case the primary focus, quite rightly, is the drama and the scientific knowledge and understanding that accompanies the main objectives. The computer here takes on a secondary role, but quickly adds a dimension to the activity that would otherwise simply not be possible in the primary school. A dramatised interpretation of a key moment in science history becomes a complete presentation in itself that can be edited, re-edited, replayed and enhanced by the inclusion of music, sound effects and titling. As with all aspects of good primary science, it can be completed entirely by the children. It is this process that is exactly what generations of educational broadcasters have been doing for many years with their schools' programming. However, the power that was once exclusively in the hands of the professional production company is now in the hands of the teacher, and more importantly, the pupil. As this technology advances, this form of recording and presentation becomes a new kind of literacy and has many opportunities for cross-curricular applications and dimensions. Ironically, it uses many of the 'traditional' literacy skills but this new relationship with these various forms of new media is a powerful new tool that we ignore at our peril (Millwood, 2000).

Bibliography

Ellis, P., 'Nobel Women', *Breakthrough* vol. 3, no. 3, PREtext, September 2001.
Medley, S., 'Budget Blockbusters', *Computeractive* issue 108, VNU, London, 2002.
Millwood, R., Chapter 4 in *ICT and Literacy*, Gamble, N. and Easingwood, N. (eds), Continuum, London, 2000.
Monteith, M., *Peer Group Editing and Redrafting*, Focus on Literacy, Micros and Primary Education (MAPE), Newman College, 1998.
Williams, J., 'Galileo after the trial: a short play', *Breakthrough* vol. 2 no. 3, PREtext, September 2000.

Chapter 9

Using ICT to present scientific investigations

Throughout this book we have continually stressed the importance of the contribution that ICT can make in supporting the key scientific investigative skills. The interactivity, provisionality, and the speed and automatic function that the computer brings can make a huge contribution to the teaching and learning of all aspects of primary science. However, ICT can also be used effectively to communicate the results of scientific enquiry to as wide a range of audiences as possible. This is of vital importance, as a complete investigation should not only consist of the experiment itself and the findings, but just as important, also show the reasons for the experimentation and give a justification of the results.

Why present scientific investigations?

A good scientific enquiry should often produce proven evidence to substantiate its findings. It is often possible to find excellent investigative science work in schools, with or without the use of ICT, but occasionally the outcomes are presented in such a way that does not do justice to the quality of the investigation. Perhaps the children have not understood completely what it is that they have done, or indeed the significance of their results. They may have produced some detailed and attractive graphs, yet have little comprehension of them, or more importantly, any understanding of why they got those particular results. In Chapter 1 we have given an example of a study of woodlice, which if done thoroughly, would give the following results:

- Highlight the characteristics of this particular creature and place it within the hierarchy of the animal kingdom;
- Identify the environment in which it lives, both *in situ* and in the laboratory;
- Show, by careful and benign scientific investigation, how this creature adapts to its environment;

- Enable children to construct and devise investigations in order to reach conclusions about the above.

All of this will require the children to come to specific conclusions at each stage, so as to enable them to progress further along in their investigation. The evidence produced at each of these stages is itself an answer to that particular part of the investigation. However, it will also provide pointers to further opportunities to investigation, and indicate in which direction these should follow.

Alternatively, children may have an excellent understanding of their investigations, but they may require more time than is available in an overloaded timetable to complete their presentations. This may be even more apparent if the science has been studied to some depth, and the presentation of the data is correspondingly complex. It seems a pity that sometimes the literacy hour could not be used to help with this – it would surely benefit both science and English. The conclusions to any investigation, regardless of the subject area, are crucial to the teaching and learning process, and just as important, these findings should be presented in a way that is relevant, coherent and detailed. The writing of these results, and the clear, concise language needed, are essential to the development of literacy skills.

Why use ICT?

The same power that ICT brings to the investigative phase of primary science can also be used to collate, refine and present the results and conclusions. If we firmly accept that the teaching of Literacy and Numeracy can be actively supported by science and vice versa, then the writing up of a report of an experiment need not necessarily mean producing simplistic work with a few drawings, geared specifically for only the teacher to read, or to be pinned on the wall in the classroom. A wider audience can be created if work is shared with other children within the school or even within other schools via the use of the Internet or email, which need not necessarily be in this country – they could be anywhere in the English-speaking world! On a more local scale, this work can also be presented as a *PowerPoint* presentation during parents' evening with a minimal amount of alteration. Any data that may have been captured and recorded, perhaps during a data-logging activity or during a plant-growing exercise can be used in many different ways to support numeracy. Whatever approach is taken, by using the power of ICT a whole host of new, original, exciting and innovative opportunities are presented. It is the purpose of this chapter to highlight how some of these opportunities might be realised using hardware and software that is cheap and readily available, or in some cases is supplied with computers at the point of purchase.

What hardware is required?

One of the most notable developments in recent years has been the increasing availability of affordable yet good-quality hardware. This not only includes computers and printers – the staple requirements of any ICT – but also other peripherals, particularly scanners and digital cameras, and as far as science is specifically concerned, electronic microscopes. These have added a completely new dimension to the amount, type and quality of work and have suddenly given ICT a very real purpose as well as the 'value-addedness' that is extremely attractive to many primary schools. The possibilities are suddenly greater; ICT is now about providing powerful tools and a range of resources that could change the ways that both children and their teachers think about teaching and learning. Primary Science, being practical and investigative by its very nature, has particularly benefited from this new equipment.

Case study – using a computer microscope

Our thanks once more to Paul Edwards and his class at Tewin Cowper Endowed Primary School, near Welwyn in Hertfordshire, who had just acquired, through a special grant, an Intel Play QX3 computer microscope. This is a relatively inexpensive microscope that can be connected to a computer and in many respects is very similar to a traditional light microscope. However, the image is displayed on the computer monitor rather than being viewed through eyepieces on the microscope itself, which ensures that the image is larger and therefore easier to see. This image can then be used in a number of flexible ways; it can be saved and inserted into a report or presentation, or it can be projected onto a screen through a data projector so that direct teaching can take place to the whole class or to groups. For further ideas one should read *Using the Digital Microscope* by Brian J. Ford (2002).

We were able to observe several groups of children from Years 5 and 6 use the microscope in their science work. Each of the groups was of mixed ability and gender. In keeping with the standard National Curriculum requirements, they had been studying the parts of a flower. They had no difficulty in mastering the ICT element of the package, but needed some tuition on how to use the actual microscope, but little more than they would need to use the more traditional light microscope. The authors have used these in their own schools, and have found that older primary children cope with them very well. The appropriate use of a microscope adds a new dimension to the science that we teach. It allows the user to see things that would otherwise be far too small to see with the naked eye, and as such is a key tool in extending scientific understanding. This might include cells or very small insects.

The children at Tewin found that by observing the parts of a flower at varying magnifications, obtaining printouts of what they could see, and generally making use of the greater flexibility that ICT allows, the science itself became more interesting and relevant. Whilst previously they might have labelled the parts in the spaces alongside a printed diagram of a standardised flower, they were now able to write their own labels against the magnified picture on the screen. Even more important, from both an educational and a scientific point of view, they were motivated to look at the parts from a variety of different flowers. They were thus able to observe that whilst most flowers may or may not have these various parts, they did not necessarily look the same. This is something that no diagram on a worksheet provides.

However, this can be taken one stage further. An additional advantage of this particular microscope and the software that accompanies it is that it enables whatever is on the slide to be recorded either as a video clip or as a still image. One great advantage that this microscope has over the light microscope, even if it did not have the added computer graphics, is that several pupils can use it and view the resulting picture at once. With the light microscope, even a small group would have to take turns, and would need to be supervised whilst doing so. With the computer microscope it is even possible to let the whole class see what is on the slide, with the minimum of supervision.

As a final exercise to explore the potential of the microscope, we were able to ask the children to observe a printed text, at all three magnifications: ×10, ×60 and ×200. This would have been difficult with a light microscope as most of the light shines through the specimen from below. The children were fascinated to see how imprecise the printing appeared to be even at only ×60 magnification. Some examples of this are illustrated in Figures 9.1 to 9.3. We suggest that this could be a part of any study of printing or how books are made. A way to bring science into the Literacy Hour perhaps!

These three images were printed out directly from the microscope by the children themselves.

Figure 9.1 Printed text at ×10 magnification.

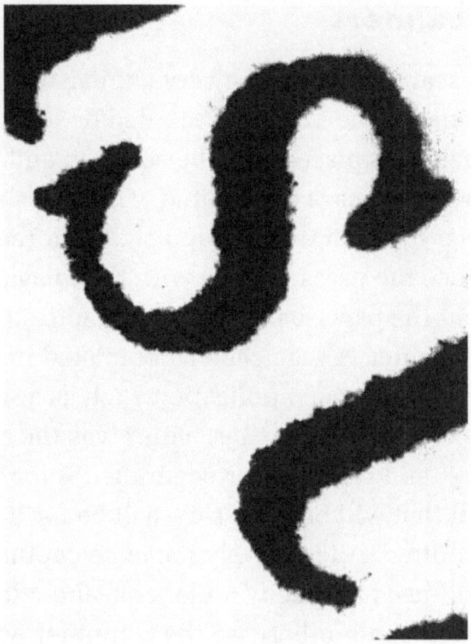

Figure 9.2 Printed text at ×60 magnification.

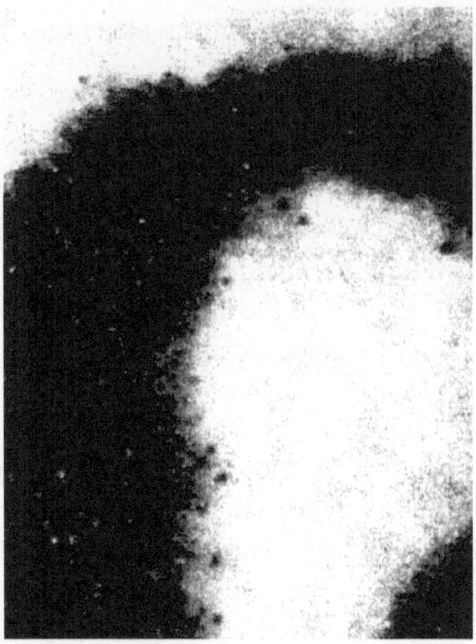

Figure 9.3 Printed text at ×200 magnification.

Scanners

A scanner works in a very similar way to a colour photocopier, only in this instance the image is recorded or scanned into a computer. The required paper picture is placed into the scanner, and when the appropriate button is pressed, a preview image is captured. This gives a preliminary image that then allows the user to select the required size, contrast and the brightness of the final image. Once the user is happy with the image the final scan is taken; this is the moment that the paper-based image is digitised and is then available for use on a computer. A scan can be completed in several different ways. It can be captured from within an application such as a word processor, desktop publisher or photo-editing package. This gives the advantage in that it is ready for immediate use, and can be inserted straight into the required position within the document. All that will be necessary will be for the image to be resized or cropped as required. Alternatively it can be captured outside of an application, but the user will need to specify a filename and a location such as the desktop or 'My Documents' otherwise the computer will not know where to save it to. The image can then be used in any number of ways. It can be saved to a floppy disk or made accessible over a network. Although a slightly more complicated solution, the latter allows greater flexibility as it means that the 'raw' image can be accessed by a wide number of other users and workstations. It is important to remember that any image or piece of text is subject to copyright laws.

Digital cameras

The ability to take a photograph, view it immediately, and then delete it if it is not deemed an appropriate record, all whilst still on the camera, is very powerful indeed. The use of a digital camera provides this option. This is a type of camera that does not use film; instead the image is saved onto a microchip inside the camera. This chip can be of varying memory capacity, but is usually of 2 or 4 megabytes. Generally speaking, a 4 megabyte chip will hold over one hundred standard quality images, or about twenty high-quality images. Standard quality images are quite suitable for the majority of uses in the primary school, for although high-quality images do give a slightly better quality of picture, they take up a vast amount of memory space and take a long time to download or transfer to other applications. Clearly, the ability to take and store over a hundred images at one time is very useful to the primary school teacher, especially if it is being used to record any fieldwork. Besides, given the quality of a standard classroom printer, the differing results will not be that apparent. The images can be transferred to a computer in a number of different ways, depending on the make and type of the camera. Some cameras are connected directly to a computer via a cable, whereas others actually take a standard floppy disk inside the camera. The image is then stored on the floppy disk, which is then removed from the camera and inserted into the floppy disk drive of the computer. A third option is where the memory card is removed from the camera and is laid into a caddy, which is the same size and shape as a floppy disk, and is then inserted into the disk drive in the usual way.

Once an image has been digitised and has been saved as a file on a computer or a network server, it can be used in many flexible ways. It can be opened within an image manipulation and editing package such as *MGI Photo Suite* and then can be altered to suit the occasion. It can be cropped, edited, have objects or people removed or even inserted, have the colours changed or special effects added. It can be rotated, tinted, or perhaps just have its file format changed. It can then be resaved any number of times in any number of formats, and can be inserted into any other document such as a report created in *Microsoft Word, Publisher, Front Page* or *PowerPoint.* Once in a file in these applications, the images can then be further cropped, resized and used to illustrate the work. One image might become a photograph on a web page, newsletter or report, or it might provide a 'watermark' as an illustration appearing within the page itself, such as in the example of the 'Flight' worksheet in Figure 7.1. A single image can be used an innumerable number of times, in a great variety of ways. This now means that even quite young children can take responsibility for the process from start to finish. They can decide what they wish to photograph, capture the image, and decide if they wish to use it, but if not then delete it. If they are going to use it, then they can download it into a computer, save it, alter it within image manipulation software, open it within an application, copy and paste it into the correct position as many times as they want and if necessary resize it and crop it. Moreover all of this can be done with a minimum of adult intervention. Prior to the invention and widespread use of the digital camera and scanner this could not have been done easily, if at all. Just think about the process that would have been involved:

1. The child would have taken the image using a camera containing film.
2. The child would have to wait until that film had been finished.
3. The film would then have been sent for processing.
4. The prints would return from processing – possibly anything up to one week later.
5. The prints would then have to be examined for suitability of use. On average only one photograph in six is deemed 'good', e.g. there are no thumbs or fingers over lenses, and the image is properly framed – the likelihood is that this is increased if a non-SLR camera is used (Single Lens Reflex, where the user sees the same image as the eventual photograph).

6. The chosen photographs would then have to be cropped by hand using scissors or a trimmer and then glued into the required position. And once this is done it is permanent.

Just think how long this would take! Also consider the other resources and difficulties that can arise. As anybody who has worked in a primary classroom knows, the moment that glue is involved the potential for disaster is ever-present! The children need to make sure that everything is correct before it is stuck down, that not too much glue is used and that there will be no need to move or remove the photograph. If there are any spelling or punctuation errors these cannot easily be changed, and even if they can, corrections can spoil the presentation. However, if the same activity is carried out using ICT, any aspect of the process can be changed at any time without impinging greatly upon any other part of the process. There may need to be some reformatting of a page in order to accommodate an altered image or an extra line of text. However, this can be fixed easily, quickly and without a trace of any alteration that it may have gone through. This is an excellent example of the provisionality of ICT, the notion that nothing is ever quite permanent in that it can always be changed, if this is what is desired. Once an image has been digitised, it can be used in all of the above ways. Through the use of a scanner, even having a traditional photograph no longer represents the end of the process as it once did, for once scanned it is digitised and as we have already seen, can be used in many further ways.

This also provides another excellent example of the computer being used as a labour-saving device. Previously a child may have drawn the equipment that was used in the investigation in order to adequately explain and illustrate what had happened. This is a very time-consuming process and relies greatly on the child's artistic skills in order to copy accurately all of the elements of the experiment. It is often very important that some forms of image are used to illustrate children's scientific investigations, but this is not the focus of the activity – that is the science itself. However, a child may have good scientific understanding but poor presentational skills, so work produced by hand may not accurately reflect the pupil's true capabilities. The use of a digital camera allows the child to record the equipment and other elements of the investigation accurately and quickly. These images can then be manipulated within the chosen application to get results that are bright, effective and correctly sized and formatted. Once again this allows the child to concentrate on higher-order skills, perhaps analysing the results in greater detail, or perhaps using the time to present these findings in a much more powerful way, perhaps through the construction of a web page or a *PowerPoint* presentation.

Collecting images from other sources

It is always preferable for children to capture their own images. They are recording their own work and they can place their own unique interpretation upon it. They should also have opportunities to extend their own understanding and ICT capability in the use of a range of appropriate hardware devices and accompanying software. However, there are other sources of images available to extend and develop the presentation of science work. There are images that are commercially available on CD-ROM or for purchase over the Internet. These have the advantage in that when purchased, the user is effectively purchasing a copyright licence so that the images can be used freely in the classroom, without fear of prosecution. This also includes clip-art, cartoon-style images that can be used to illustrate documents. These can be found from several different sources.

There are many web sites that contain freely downloadable clip-art images, but these need to be found via the use of a search engine such as Google (www.google.com) or Yahoo! (www.yahoo.com). They are also available on CD-ROMs, usually reasonably priced or can be available as freeware (software that is supplied for free, often with magazines) or shareware (where a fee is payable after a trial period, usually thirty days). However, one of the easiest and most flexible means of inserting clip-art, especially if a *Microsoft Office* product is being used, is the use of *Microsoft Clip Art*. This comes on the Microsoft Office installation CD-ROM and can be installed for use with *Word, PowerPoint, Excel* and *Access*. It can also be used with *Publisher* and *Front Page*, which do not come as standard on the *Office* disk. However, there are benefits in not installing it. For example, it takes up a large amount of space on the host computer's hard disk and it then becomes difficult to update. By following the Picture – Clip Art link from the Insert menu, it is possible to download clip-art directly from the *Microsoft Clip Art* gallery, which is online. This has the added advantage in that new clip-art is being added all the time, so the user has access to a huge library of clip-art without the difficulties of having it installed on a computer. There is a powerful search function, and the site is easy to navigate, so this ensures that clip-art can be accessed quickly and easily. It is free to access, and the image can be inserted straight into the document that is being created, where it can then be resized. It should be remembered, however, that there are limitations. Clip-art cannot be rotated from within most applications, so that if for example a character is facing the wrong direction, then this cannot easily be corrected. Also, as this is *Microsoft Clip Art*, it is not directly compatible with products from other software companies. Therefore it may be difficult or impossible to download and cannot be resized without appearing to go out of focus, and cannot often be found as easily identifiable separate files. However, if Microsoft products are being used, then this represents a simple and powerful way of enhancing work via the Internet.

Using ICT to present science by 'traditional means'

One of the most popular means of communicating the results of science investigations is by the use of a word processing package. This enables the user to ensure that their work is presented neatly, carefully and colourfully. The work can look attractive through the use of different font styles, sizes and colours, and these can be effectively used to stress key points and ideas. A range of attractive borders can also be included to help to improve overall presentation. The work can be amended, edited or altered any number of times without leaving any trace of the number of revisions that it may have gone through. There will be no rubbings out, or correction fluid on the final piece of work. It can be stored electronically and printed as required, so whatever the need or purpose, a fresh copy can always be obtained, months or even years after it was originally produced. A piece of work can evolve over a long or short period of time – it is this provisionality that previously was unavailable to the teacher and learner. This is a great source of motivation to all children, but particularly for those who find writing difficult. The user can also enhance and embellish their work by including images in the form of clip-art or photographs from a range of sources including a digital camera, scanner or the Internet.

Case study – a use of simple word processing to enhance the presentation of a science topic

Our thanks to Bryan Anderson and his Key Stage 1 class of children at St. Christopher School, Letchworth, Hertfordshire. The children were carrying out an integrated project based on the ancient belief of the four elements – earth, air, fire and water. The project incorporated many of the subjects of the curriculum and the science involved such aspects as animals in the sea, and the nature and causes of fire. As part of their practical work the children made windmills when studying air and built lava lamps when learning about water.

Figure 9.4 shows word processed instructions describing how to make a lava lamp, as written by one of the children.

For recording much of their written work the children used *Microsoft Word* and they used the Internet to obtain general background information. Figure 9.5 shows an acrostic poem about fire, word processed by a child in the class.

Lava Lamps

You will need
Salt.
Water.
Jar.
Colouring.
Oil.

1. Put the water in the jar.
2. Put in the oil.
3. Put in the salt.
4. Watch the result.

The oil will not mix with water because the water is lighter.
The salt pulls the oil down to the bottom.
The salt dissolves and the oil pulls the salt back to the surface.

By Jenny

Figure 9.4 Instructions describing how to make a lava lamp.

Fire flickers

Flickering dancing burning shining.
It shimmers and licks all the time.
Red embers in the fireplace.
Embers shining after.
Glowing in the fireplace.
Living when you make it.
October the 16th is when you start making a fire.
Water puts it out.

By Jenny

Figure 9.5 An acrostic poem about fire.

Using ICT to present science by 'non-traditional means'

Important though it is, there is much more to the use of ICT than simply as a means of ensuring high standards of presentation. As we have already discussed in this book, not only is it a tool for learning, it can also change the way that children learn and the way that teachers teach. ICT provides a dynamic environment that offers the user many hitherto undreamt of opportunities, and places them firmly in the hands of children and their teachers. This suggests that teachers and subsequently their pupils may have to change their perceptions about the ways that work is presented. This should lead to an understanding that work does not necessarily have to be printed out to be deemed as complete, or even appear as a traditional paper 'page'. Indeed, with the hyperlink facility in *Microsoft Office* applications, any piece of work can be linked directly to any other named file either on that computer or on the Internet, just by pointing and clicking. Therefore children can produce a report of their science investigations and then can directly link this to another file. This could be another page in a *PowerPoint* presentation by the same authors, another presentation by someone else in the same class, or indeed presentations by anybody else anywhere in the world, as long as it is online. This is an extraordinarily powerful idea! There could also be links to web sites where further supporting information is displayed, so a hyperlinked option to the effect of 'click here to find out more information about dinosaurs' could link directly to the Natural History Museum's website – or indeed any relevant web site in the world. This is a truly exciting opportunity.

One of the many advantages of using ICT to present any work to an audience is the fact that it can transpose existing information into a number of different formats. A piece of word processed text can be transformed into a web page, a newsletter or a presentation easily and quickly. For example, *Microsoft Office* application software has an option that allows any work created within it to be saved in HTML format – in other words it is ready to be posted onto the Internet immediately. Some users often prefer to use a dedicated web-authoring package such as *Front Page* or *Dreamweaver* and again, this can be easily achieved by copying and pasting between applications. This allows for all kinds of possibilities. Children can complete their investigation, write it up and illustrate it using ICT and can then post it on their web site. This can then be viewed from anywhere else in the world and if they have a message board facility, others can then send in comments about the work. It might be something as simple as someone saying how much they liked their work, or it could be another school explaining what results they got when they carried out a similar investigation. It might be seen by specialist scientists in a university or a research department in a company or hospital somewhere, and as was discussed in the multimedia chapter, can stimulate a wide range of discussion, possibly involving experts in a particular field. The use of email can also play a significant role here, as pupils can send work to partner schools or even universities and colleges to invite comment from different sources. The responses can then be fed back into their own work at any stage, and might even provide data for comparative studies across several different schools. This would also provide good opportunities to link together clusters of primary schools, perhaps to their local secondary school, sixth form college and university. The Internet becomes not only a source of information which children can use themselves for research purposes, but in turn becomes a place where they can add to a body of knowledge and provide a source of information that others can use.

Children should rarely be encouraged to simply copy onto the computer work that they have already carefully written out in long hand. This may well have been corrected in its rough state, and then beautifully presented, but that should be the end of it. This copy-typing has always been a pointless, time-wasting activity as neither the computer nor the children are remotely extended. Learning is not taking place as the power and potential of the computer, or for that matter the user, are not being fully employed. The opportunities now afforded by ICT as described above, and especially the use of the Internet and hyperlinking within and between applications, ensures that there is great potential to really extend the learner. With the vastly increased levels of access to ICT in the primary school in the last few years, there is now the opportunity to let children compose their work on the screen. This composition no longer needs to be only a linear, pencil-and-paper type of writing. Scope now exists to create something that can be produced with a computer. This in turn means that the user now has other considerations. If the work is to be displayed on the World Wide Web, then clearly the user is now engaged in a publishing activity, and as a result of this, children have to consider web design. Of course it is perfectly acceptable for them to have notes and their raw data to assist them, but the need to synthesise the information, lay it out on the screen, draw pictures using the drawing tools, add images, tables, titles and borders, are all skills that need to be developed. In this context they become skills not only of ICT, but also of science, used to communicate scientific discovery, with the help of new technologies, to a wide range of audiences. So the child no longer has to consider just content, but also the way that it is to be presented. Previously a hand-written science report would have consisted of some text explaining in detail the experiment, as well as perhaps some drawings of the equipment that was used. On completion this would then be placed on the wall as part of a display, or would perhaps disappear into a topic book, viewed by the teacher and at most a few other children and perhaps some parents on open evening. However, when using ICT the child now needs to consider a number of other aspects, depending on how the work is to be presented. This will include a sense of audience, and a feeling for constructing a piece of work that need not necessarily follow in a linear form. An individual's piece of work can be linked to the work of other children or to commercially produced web pages. This is a valuable feature because it enables the children to connect their work with relevant research or further reading. A study on minibeasts could link to the relevant web pages of the Natural History Museum's excellent web site, or a topic on flight could link to the Science Museum. So the children are not only considering their own findings; they are inviting their audiences, whoever they may be, to participate by following given links and making logical connections between the content from the writer to the 'expert' content from an external agency.

Creating a presentation

We have already discussed in some depth the advantages that using ICT can bring to the presentation of the findings of scientific investigation. These can be the ability to incorporate images from cameras, scanners and other electronic sources such as CD-ROMs or from the Internet. To incorporate sound, tables and text; and the ability to hyperlink to other files or documents on both local and distant computers. However, the use of presentation software allows all of these different aspects to be used and integrated with greater flexibility. The ability to create different presentational slides easily and quickly, and to link these together in a number of different ways, complete with illustrations, is a very valuable asset. Add to this the ability to insert video clips and sounds, both of which can be recorded by the children themselves, as discussed in the chapter concerning ICT and Drama, and the children will have complete control of their own learning. This represents an entirely new way of illustrating the results of their scientific investigations. With a data projector, the children can project their work onto a large screen and can talk about their work, rather than just write it down using traditional pencil-and-paper methods. It has always been a part of primary school life for children to share their work with others in this way, perhaps during a class assembly or open evening, but the power that presentation software brings allows a new dimension. It is bright, attractive and colourful, and allows a multimedia approach simply and easily. It immediately attracts the viewers' attention and holds their concentration. All of the required features can be easily accessed by pointing and clicking 'hot spots' on the screen; there is no longer the need to use separate slide or movie projectors, tape recorders or CD players, as all can be controlled through the software.

A sense of audience, a sense of pride

However, there is more to this than presentational convenience. The real advantages of using ICT are embedded in what it can bring to teaching and learning, and particularly, how it could connect to literacy. In any type of presentation or report, the writer has to consider content and to a certain extent layout, but now the viewer also needs to be borne in mind. The need to write succinctly through the use of bullet points encourages the author to prepare a presentation that not only incorporates all of the above features, but also makes the writer engage in the higher-order literacy skills needed to include the audience who are viewing it. Careful consideration needs to be given to key features such as background design and colour so that it does not detract from the text. If animations are used they must help to reinforce a point rather than detract from it. There is nothing worse than a presentation that has too many animations flying around the screen or is difficult to read. A well-structured and composed presentation can really bring that 'value-added' component that we have mentioned frequently throughout this book, as during the presentation the demands are made on the viewer rather than the writer. A presentation can be used in several main ways; it may be used, as previously suggested with a data projector to a group in a class, assembly or open evening; it can be left on screen for other children to access, perhaps working round an ICT suite where there may be a dozen or so different presentations on screen. Of course, the hyperlink and network capabilities and functions, can be linked together! The presentation can also be run on a stand-alone machine, but can be set up to 'loop' continuously until the Escape button is pressed. This might have a spoken commentary recorded onto it, so that the writer doesn't need to be present at all. If this is then subsequently placed online, or is emailed to a partner school, then a new dimension to the science report immediately becomes apparent.

Although most primary schools currently have access to powerful presentation software, and it is preferable to use this where possible, there are nevertheless other ways of creating and displaying presentations through the use of a web browser. It is not always appreciated that this type of software is extremely versatile for a whole range of uses, quite apart from displaying pages from the web. It of course comes as standard with every new computer. By working offline, so that the browser does not constantly try to load fresh pages, it is possible to view a whole host of different types of documents and images easily and quickly, simply by dragging and dropping the files to be viewed into the browser window. These can then be viewed by clicking the forward and back buttons on the browser. For example, if the user wanted to view and show a series of images they would launch their chosen browser, stop the home page from loading and then click on the 'Work Offline' option. The user then needs to click on the 'Restore' button on the top right-hand corner of the screen so that the desktop can be seen beneath. The user then finds the file image that is to be viewed, and this and any other images are then individually dragged and dropped into the viewer. This has to be done in the correct order, i.e. the order in which they are to be viewed. By clicking the forward and back buttons the chosen images can be seen on the screen without the need to be loaded into another application first, which is very useful for quick demonstrations or explanations. Alternatively, the required file can be opened via the file menu and 'open' selected, and then chosen in the usual way.

But what about the teacher?

Although the focus of this chapter has been about how children can use ICT to present the results of their scientific investigations, it might seem pertinent at this point to illustrate how teachers can use this same potential to enhance their teaching of primary science. Although wherever possible children should be encouraged to undertake practical experimentation, the teacher may need to use presentation software to demonstrate and explain a range of scientific concepts at an appropriate time during the children's project. These might otherwise be too difficult to illustrate due to constraints of time, size, cost or safety. Examples of these might include the growth of a plant (too slow), cell division or replication (too small), the water cycle (too slow and too large) or nuclear fission (too dangerous)! Although it is true to say that educational broadcasts have fulfilled this requirement for many years, the use of ICT brings this power to teachers for the first time as they can now create their own electronic presentations.

Some examples of these presentations can be found by visiting www. …

Reference

Ford, Brian J., *Using the Digital Microscope*, Rothay House, Cambridge, 2002.

Index

2Count 40
2Simple Video Toolbox software 38–40

Ager, R. 52
animal types case study 61–6
Apple Macintosh 36–7, 137
art galleries 119, 125, 127
Ask Jeeves 50
Atkin, R. 43, 45
Atkins, P. 54

Black Cat Toolbox 34, 51, 56, 73
British Educational Communications and Technology Agency (BECTa) 7, 34–6, 39

case studies, control technology 109–16; data logging 88–91; database 60–8; hardware benefits 145–7; ICT 11–14; presentation of science 152–3; spreadsheet 75–9
CD-ROMs 2, 5, 24, 36–7, 38, 40, 50, 53, 127, 131–2, 133, 138, 141, 142, 151–2
Channel Four 12
children, and copy-typing 155; and effective use of computers 27–8; extending thinking though focused questioning 91–2; gender aspects 114; as individuals 27; interactivity of 26; as learners 27; in pairs 27; as peer tutors 27; *see also* case studies
Claris Home Page 44
CoCo 114
computers, access to 54; additional equipment/software 24–5; clarity of screen 39; effective use of 27–8; how/why they are used 18; and Internet, web facilities, email 26; networking facilities 26; power of 21; and sort/display of data 19; time/access to 27; use/misuse of 32–3; as value-added component 11, 18–19; video equipment 138; *see also* control technology
control technology 85; benefits/importance of 116–17; and buying kits 96–7; case studies 109–16; and construction of gear trains 100–7; and cross-curricular experience 99–100, 116; described 96–100; and hierarchy of ideas/decisions 108; and input/output requirements 98; and making models 96–8; as overlooked area 98–9, 107; process/progress in 108–9; teaching of 96–7, 98–9, 108; and use of questioning 100; using/developing 96, 107–9, 116; *see also* technology
Crompton, R. 22

Data Harvest 85
data logging 2; activities 85–8; case study 88–91; described 81; and developing higher order thinking 82–3; and focused questioning 91–2; and health and safety 93; place of in curriculum 81–2; power of 94; and temperature, light intensity, noise level measurement 87–8; and use of light sensors 86–7; weather project example 83–5
databases 2; branching/binary tree 25, 50, 53, 54–5; case studies 60–8; difference with spreadsheet 72; and display,

search, sort information 51–3, 72; early limitations 49–50; free text 50; power/flexibility of 49, 51–4, 68; programmes for 51; random access 50, 53–4, 56–60; structure of 51, 56; types of 50–1; use of 68–9
de Bono, E. 1
Design and Technology 95–6
digital cameras 4, 24, 148–50
digital video, organising/managing 139–41; teaching/learning considerations 141–2; technical considerations 138; use of 137–8
drama 156; areas of study 135–6; equipment needed 138; making science meaningful 136–7; role of in curriculum 137; teaching/learning considerations 141–2; and use of digital video 137–8

Early Learning Goals 52
ecological case studies 88–91

Feasey, R. 71, 73, 81
First Workshop 51, 56
Flight 121, 122–4, 156
Forces 120, 121, 133
Ford, B.J. 145
Fowler, P. 133

Gallear, B. 71, 73, 81
gardens 12, 119
gear train construction 100–7
Ghost Train 115–16
Google 50, 151
Granada Learning 34

hands-on science 14, 127–8; Archimedes' screw 129–31; levers 128–9; pulleys 129
Harlen, W. 82
Harwood, P. 81
health and safety 93
higher order thinking 82–3
Howzat! 12

ICT (information communication technology), adoption/availability of 4–5; application/contribution of 5–6; background 1–2; case study 11–14; hardware requirements 145; importance of 143; introduction/use of 2–3; link with science 8–9, 14–18; organisation/management of 25–30; place/purpose of 4–5; practicalities of 2–3; and preparation/support of lesson 19–25; reasons for 7–9, 144–5; role of 18; suites 4, 24, 26, 29–30; as tool for learning 6–7; value-added component 145; within educational establishments 127
iMac 36–7, 137
images 151–2
iMovie 137
information *see* databases; spreadsheets
Information Workshop 51, 56, 62, 63, 67–8, 69, 72
Intel Play QX3 145
Internet 4, 24, 121–5, 155

Kemmis, S. 43, 45

ladybird floor turtle case study 109–12
Lego Mindstorms 97, 112
lessons, assessment 13, 21–3; cross-curricular links 12, 14, 23–4; development of 13, 20; differentiation 13, 20–1; follow-up 14; and interactivity 26; objectives of 13, 19–20; organising/managing 25–30; planning 13–14, 19–25; presentation of results 14–17; prior learning 13, 20; resources 14, 24–5; science as key objective 30; and teacher focus 26; teaching points 13, 20
Local Education Authorities (LEAs) 33, 26

Mackenzie, J. 54
Macromedia Dreamweaver 44, 154
Mann, P. 22
'Mary Rose' software programme 32
MGI *Cinematic* 137
microelectronics 95
microscope 4, 24, 145–7
Microsoft Access 151
Microsoft Clip Art 151
Microsoft Excel 34
Microsoft Front Page 44, 149, 151, 154
Microsoft Office 34, 151, 154

Microsoft PowerPoint 46, 140, 144, 149, 150, 154
Microsoft Publisher 34, 149, 151
Microsoft Windows 34, 36
Microsoft Word 34, 149, 151, 152
Millenium Bridge 96
Millwood, R. 142
Monteith, M. 139
multimedia 131, 132, 156
museums 12, 119, 120, 121, 124–5, 127–8
My World 67, 69

National Curriculum 1, 3, 6–7, 18, 35, 92, 95, 119, 126, 133, 145
National Grid for Learning (NGfl) 4, 125–6
Netscape Composer 44
New Opportunities Fund (NOF) 3
Number Box 34, 38–40, 73, 74
Number Magic 76

OFSTED 2, 107
Ourselves case study 66–8, 75–9

Porter, J. 81
presentation of investigations, audience/pride 157–8; by non-traditional means 154–6; by traditional means 152–3; case studies 145–7, 152–3; collecting images from CD-ROMs/web sites 151–2; creating 156; digital cameras 148–50; hardware requirements 145; reasons for 143–4; role of teachers 158; scanners 148; and use of ICT 144–5

Qualification Curriculum Authority (QCA) 20, 126
Quick Time 37

Robolab software 97, 112–16

scanners 4, 148
Schemes of Work 126
science, and asking focused questions 91–2; as core subject/key objective 1, 21, 30; and data handling 79; experimental approach 4; history of 135–6, 137; link with ICT 8–9; links with technology 95; making it meaningful 136–7; as practical/hands-on subject 25, 127–31; presentation by traditional means 152–3; rate of progress in 8; reasons for investigation 143–4; skills/processes of 3–4; teaching of 2; technological element of 3; time alloted to teaching 3; topic approach 119–20
software, as adjunct to practical science work 46; bundled 33; choosing 31, 40–2; as conjectural 43–4; educational considerations 38–40; good 36; graphics/sound considerations 37; historical context 32–4; installing 37; as instructional 43; interactive component 39; language differences 39, 40; limitations of 32–3, 49–50; manuals 39; multimedia 132; open-ended use 44; options facility 39; packages 24–5; place/purpose of in curriculum 42–5, 46–7; quality of 38–40; range of 31; relevance of 31–2; requirements for 35–6; as revelatory 43; selection by hardware company/LEA 33; teacher considerations 34–6, 45–6; technical considerations 36–7; writing of 34
spreadsheets 2, 18; advantages 71; case study 75–9; described 71–2; difference with database 72; manipulate/model numbers 72; and mathematical calculations 73; when used 72–5, 79
Stephenson, P. 108, 116

teachers, and asking questions 26; confidence of 33; importance of 11, 17–18; role of 158; and software evaluation 34–6, 42, 45–6
teacherxpress.com 126
technology 95–6; *see also* control technology

Virtual Teacher Centre (VTC) 126–7
visits 12; accessibility of place 119; benefits of 119–20; educational application of 119; and hands-on science 127–31; links to curriculum 120, 127; organising/managing 120–1; and preparation of material for

120–1; and using Internet as support 121–5; and using multimedia backup 131–2

Wake, B. 54
weather sensing project 83–5
web sites 7, 34–6, 37, 40, 121, 124–5, 126, 133–4, 154
Williams, J. 83, 136
woodlice case study 11–17, 18, 19
word processing packages 34, 152–3
World Wide Web 37, 50, 53, 125, 141, 155
Wright, E. 43, 45
Write Away! 34

Yahoo! 50, 151

zoos 12, 119